Octavian Iordache
Evolvable Designs of Experiments

Related Titles

Hinkelmann, K., Kempthorne, O.

Design and Analysis of Experiments

Volume 1: Introduction to Experimental Design

2008
ISBN: 978-0-471-72756-9

Hinkelmann, K., Kempthorne, O.

Design and Analysis of Experiments

Volume 2: Advanced Experimental Design

2005
ISBN: 978-0-471-55177-5

Lazic, Z. R.

Design of Experiments in Chemical Engineering

A Practical Guide

2004
ISBN: 978-3-527-31142-2

Mason, R. L., Gunst, R. F., Hess, J. L.

Statistical Design and Analysis of Experiments

With Applications to Engineering and Science

Hardcover
ISBN: 978-0-471-37216-5

Wu, C. F. J., Hamada, M.

Experiments

Planning, Analysis, and Parameter Design Optimization

2000
ISBN: 978-0-471-25511-6

Octavian Iordache

Evolvable Designs of Experiments

Applications for Circuits

WILEY-
VCH

WILEY-VCH Verlag GmbH & Co. KGaA

The Author

Professor Octavian Iordache
Polystochastic
3205 Pitfield Boulevard
Montreal, QC H4S 1H3
Canada

1005617613

Library of Congress Card No.: applied for

British Library Cataloguing-in-Publication Data
A catalogue record for this book is available from the British Library.

Bibliographic information published by the Deutsche Nationalbibliothek
The Deutsche Nationalbibliothek lists this publication in the Deutsche Nationalbibliografie; detailed bibliographic data are available on the Internet at http://dnb.d-nb.de.

Typesetting   Thomson Digital, Noida, India
Printing   betz-druck GmbH, Darmstadt
Binding   Litges & Dopf GmbH, Heppenheim

Printed in the Federal Republic of Germany
Printed on acid-free paper

ISBN: 978-3-527-32424-8

Contents

Evolvable Designs of Experiments: Applications for Circuits. Octavian Iordache
Copyright © 2009 WILEY-VCH Verlag GmbH & Co. KGaA, Weinheim
ISBN: 978-3-527-32424-8

Preface

Several industries are confronted with the advent of complexity in terms of product and market. Complex systems are defined as structures and processes involving nonlinear interactions among many components and displaying emergent properties. Emergence is linked to evolvability for the systems of industrial interest that are studied here.

Evolvability is not biological but should be considered in the sense that the system has, to some extent, generic characteristics that are usually associated with the living systems.

The challenge of complexity imposes modifications of basic methods on the industry such as design and control of processes, testing strategy, problem solving, quality, reliability analysis, and manufacturing methods. Finally, complexity forces both the product and the enterprise to change.

Evolvability characterizes and permits to confront and break the complexity barrier. Evolvable designs of experiments (EDOEs) are presented as a constructivist approach to meet the requirements imposed by complexity on the industry. The key mathematical tools for EDOE modeling are the polystochastic models (PSMs). The term "polystochastic" was coined to describe the systems that emerge when several component stochastic processes going on at different conditioning levels are able to interplay with each other, resulting in a process that is structurally different from the component processes (Iordache, 1987). EDOEs are networks of component designs of experiments that result from the self-referencing autonomous closure between the rules of the designs of experiments (DOEs) and the dynamics of the real physical organization. EDOEs are hierarchically organized according to informational criteria in several conditioning levels of component DOE surrounding the central DOE. EDOE integrates the recursive hierarchical net organization with cyclic interactions. It combines the forward search, from the center, which opens the study of the problem, with the backward search based on real-field evaluations moving toward the center, closing and solving the problem.

The printed circuit is the industrial product chosen to illustrate the theoretical constructivist approach presented here. An outline of the printed circuit boards (PCBs) technology opens the first chapter of this book. Part One reviews different paradigms of quality and reliability in manufacturing. New paradigms such as self-adaptability, proactiveness, and evolvability appear as necessary developments.

Evolvable Designs of Experiments: Applications for Circuits. Octavian Iordache
Copyright © 2009 WILEY-VCH Verlag GmbH & Co. KGaA, Weinheim
ISBN: 978-3-527-32424-8

Part Two introduces theoretical and practical aspects of the EDOE method. Relevant concepts such as the polystochastic model, the first-order wave equation generating EDOE, and the informational approach are discussed in Part Two. Part Three presents case studies. Applications detailed in this chapter refer to PCB solderability, reliability, drilling, surface finish, direct plate, and plating voids. PCB assembly, PCBA issues are among the focused problems.

Part Four relates EDOE concepts to real devices, in this case the evolvable circuits.

Complexity imposes, and EDOE methodology enables, the replacement of conventional fixed circuits by evolvable ones. The self-adaptive devices are supposed to be able to learn and to perform elementary tasks specific to living systems. Proactive and evolvable circuits should go beyond learning, turning into autonomous and innovative devices, able to take control of and use unexpected events and opportunities of their changing environment. The major role of evolvability for the emergent manufacturing systems is emphasized.

Undoubtedly, EDOE methodology is not limited to circuit manufacturing.

It is hoped that the case studies presented in this book will help readers to use EDOE methods for their own applications. EDOE specifies how to break the complexity barrier by changing the done and fixed design of experiments, technologies, or devices by constructed and evolving ones. The book will be useful to industrial engineers, scientists, and entrepreneurs from different production domains, to students of systems science, engineering, computer science, statistics, and applied mathematics.

Montreal, January 2009   *Octavian Iordache*

Abbreviations

A/A	air-to-air
ANOVA	analysis of variances
ASF	ampere per square feet
AST	accelerated stress testing
BR	bacteriorhodopsine
CTB	cell test board
DI	dissociation–integration
DNA	deoxyribonucleic acid
DOE	design of experiment
DP	direct plate
EC	evolvable circuit
ECF	electrochemical filament
EDOE	evolvable design of experiment
ENIG	electroless nickel–immersion gold
FHS	finite hole size
FMEA	failure mode and effect analysis
FPGA	field programmable gate array
GF(m)	Galois field
I/L	inner layer
IST	interconnect stress testing
L	Latin square matrix
L/L	liquid-to-liquid
MECS	microenergy chemical system
MEMS	microelectromechanical system
NN	neural network
PAC	proactive circuit
PCB	printed circuit board
PCBA	printed circuit board assembly
PSM	polystochastic model
PTH	plated through hole
RMS	reliability management system
RNA	ribonucleic acid
RSCC	random system with complete connections

Evolvable Designs of Experiments: Applications for Circuits. Octavian Iordache
Copyright © 2009 WILEY-VCH Verlag GmbH & Co. KGaA, Weinheim
ISBN: 978-3-527-32424-8

SAC	self-adaptive circuit
SEM	scanning electronic microscopy
SKUP	quadruple: *S*-states, *K*-conditions, *U*-operators, *P*-possibilities
SM	solder mask
SPC	statistical process control
TMC	test minicoupon
TQM	total quality management
W	Walsh matrix
WE	wave equation
WH	Walsh–Hadamard matrix
WP	Walsh–Paley matrix

Part One
Introduction

1
Printed Circuits

1.1
Technology Presentation

Printed circuit boards (PCBs) and printed circuit board assembly (PCBA) case studies serve to illustrate the new design of experiment (DOE) methodology discussed in this book. The PCB is an essential part of the electronic circuit packaging system that interconnects the electronic components for specific tasks. The PCB provides the mechanical support and the necessary connections between the components attached.

The modern PCBs should be smaller, highly integrated, and should have faster operating speed, higher power ranges, and higher reliability.

PCBs are categorized in several ways according to

- Layer count
- Substrate
- Additive or subtractive technology
- Rigidity or flexibility.

According to layer count, the PCBs are classified into three main categories:

- Multilayer PCB
- Double-sided PCB
- Single-sided PCB.

PCBs are also categorized by substrates or base materials into three classes:

- Rigid PCB
- Flexible
- Rigid–flex.

Rigid PCBs are the most common type of PCBs especially when used to interconnect components. Flexible circuits are manufactured on polyimide and polyester substrates that remain flexible at finished thickness. They allow 3D movements. Rigid–flex boards are assembly of rigid and flexible boards laminated together during the manufacturing process.

Evolvable Designs of Experiments: Applications for Circuits. Octavian Iordache
Copyright © 2009 WILEY-VCH Verlag GmbH & Co. KGaA, Weinheim
ISBN: 978-3-527-32424-8

The most common manufacturing method is based on subtractive processing in which the metal is selectively removed from a PCB, and what remains forms the conductive circuit.

Additive processing refers to a process whereby the circuit is formed by selectively plating metal on a substrate to create a circuit layer. Hybrid methods referred to as partially additive and semiadditive are essentially subtractive methods.

The main constituents of a standard PCB are the copper foil, electroplated on titanium or stainless steel, and the dielectric prepreg, consisting of resin, which may or may not be reinforced with glass fibers, woven or nonwoven, or filaments, or other inert fillers. The prepreg is manufactured by permeating woven glass fabric with a solution of epoxy resin and then passing it through a heat treatment that removes the solvent and partially cures the resin, taking it from the nonreacted stage to the "B-stage," partially cured. Prepreg, that is, "B-stage" with different fabric weaves, different resin systems, and different resin/glass ratios, is accessible. Despite their variety, all the PCBs are basically composed of conductors, dielectrics, and vias. This generic structure determines the few basic steps that are common to most PCB fabrications: materials preparation, inner-layer processing, laminate preparation and lamination, drilling, making conductive holes, imaging, developing, electroplating, etching, solder mask (SM) application, surface finish application, and routing and testing (Coombs, 1996).

1.2
Inner-Layer Processing

In inner-layer processing stage, each layer is processed in a printed circuit structure by resist or film application, imaging, and developing, followed by copper etching and resist stripping. To ensure adhesion between the layers and the additional prepreg layers, all layers are chemically treated by oxidation, to black or red oxide, by application of thin coats of metallic base with bonding properties, such as tin complex compounds or even by the creation of specific topography by inhibited acid treatment.

1.3
Materials Preparation

The core material is sheared to panel size and then cleaned mechanically, chemically, or by combination of both.

Mechanical scrubbing methods include abrasive brush scrubbing and aluminum oxide scrubbing. Brush scrubbing removes a thin layer of surface copper but can produce a surface noncompatible with fine-line circuit design. Aluminum oxide produces a favorable surface for photoresist application.

Chemical cleaning is accomplished in spray chambers with agents such as potassium persulfates. Supplementary steps may include mild oxidizers.

1.4
Lamination

The lamination process involves two distinct yet linked operations:

- Layup
- Lamination.

The layup is referred to as "building up the book." The material is chosen and sized taking into account the expected lifetime of the assembled board.

During the lamination process, the thin-core inner layers are subjected to heat and pressure and compressed into a laminated panel. Prepreg or B-stage sheets are slipped between the layers to bind the layers together.

In this stage, the objective is to form the layup consisting of sheets of copper foil separated by two or more plies of prepreg. This brings the resin to a new stage, sometimes designated as the "C-stage," which corresponds to a more complete cure. The lamination is done under vacuum to remove volatiles as the B-stage cures. The resulting raw board is cleaned and sized.

Sequential lamination is a technology that takes several multilayered circuits and laminates them together to produce one or more multilayered boards. In sequential lamination technology, the panels are drilled and another lamination step takes place for the outer layers. When a design includes different types of vias, it typically requires a set of sequential lamination and electroplating cycles.

1.5
Drilling

The purpose of through-hole drilling PCB is twofold:

- To produce an opening through the board that will permit a subsequent process to form an electrical connection between top, bottom, and internal conductor pathways.
- To permit through the board component mounting with structural integrity and precision of location.

There are critical points to take into account for drilling operation: the alignment between the inner layers and between the inner and outer layers, the drill geometry correlation, speed, and material with smear formation control, the sidewall's integrity, and the removal of residue. The multilayered board may include different types of vias, for instance,

- Through holes
- Buried via
- Blind via.

Buried via is drilled through inner layers and does not exit to either outer layer. Blind via starts at one surface layer but terminates before it penetrates all layers. The blind via is a through hole connecting the surface to one of the internal layers. For some electrical connections, the mechanical drilling was replaced by laser ablation, allowing different sizes of blind via holes in the external layers of the board. Adequate cleaning methods such as mechanical, chemical, or plasma etching follow the drilling. A nonconventional method to make well-defined blind vias in outer layers is to apply the image on the external monomer layer, the photodefinable dielectric that is polymerized by exposure and by blind via emptied as a result of developing. Such materials can be made conductive using specific methods.

The drilling or ablation causes smearing or residues of epoxy resin on the inside of the holes. Aggressive chemicals remove these imperfections.

Hole cleaning refers to a process called desmear and to the closely related process of etchback. Desmear removes the melted resin smears that result from the friction of the drill cutting through the material. At present, the most widely used chemistry is sodium or potassium permanganate.

During etchback, glass fibers are etched in addition to removing the resin smear. Glass etchings include hydrochloric acid. Etchback with plasma can be achieved by varying the type and the amount of reactive gases.

1.6
Making the Hole Conductive

To provide the intended interconnection between layers, the holes must be coated or plated with a conductor substance. Therefore, the next stage is to render all drilled surfaces conductive. The PCB substrate is not conductive, so a nonelectrolytic deposition method is required.

The hole is made conductive using copper electroless or direct plate. The electroless technology consists in plating a thin copper layer with a controlled structure thickness and adhesion on each material of composite complex.

The main steps of electroless process are

- Cleaning the hole and surface to ensure that the copper is not deposited on the base material without a passive interface.
- Activation and acceleration by depositing a catalyst on the hole surface that allows subsequent copper deposition.
- Copper deposition, that is, reduction of the copper ion in solution causing copper metal to deposit on the hole surface.

The controlled structure and thickness will ensure the film's continuity after surface preparation and before electroplating. The direct plate allows avoiding the electroless or electroplating voids or cavities. The good adhesion will ensure the removal of one of the major defects affecting the product quality and reliability, the inner-layer separation.

In addition to the process of electroless copper deposition, a process called "fully additive" can be used to build up the conductors without electroplating processes.

1.7
Imaging

Typically, the imaging process includes three steps:

- Photoresist application
- Exposing or printing
- Developing.

Photoresists are available in the form of dry films and liquid resists.

Film photoresist is applied with heat and pressure to the surface of the panel. This is done in a laminator.

The printing of the image is usually accomplished by placing a film or glass phototool between the panel to image and a light source. Advances in resist chemistry and laser tools have made possible "direct imaging" that expose images on the photoresist directly from computer database design information and does not need a phototool, an important option for special applications.

The selective copper plating allowing the buildup of the circuit is possible only by exposing the conductive surfaces according to predetermined design. The rest of the base copper foil is used only temporarily to conduct the electrical current for electroplating and is for that reason protected. If there are several options for inner layers, the image for outer layers is defined by means of a dry film.

For good adhesion, the copper base, electroless copper, or electrolytic copper for panel plate technique should be textured to an appropriate roughness by using a mechanical wet abrasion product, such as aluminum oxide, or pumice, together with an acid cleaning. The film lamination depends on the temperature and on the pressure of rollers, and advances according to both chemical composition of the monomer film and its thickness. The exposure step, the polymerization of the film on nonuseful base copper surfaces, is a difficult operation due to the degree of precision imposed. Misalignment may have a significant impact on the PCB quality. The cleanliness of the environment and a perfect contact between the board and the pattern image, ensuring the perfect vacuum, are critical. The accuracy of the energy application defines the resist wall quality.

A distinctive imaging method for ultrafine lines is based on the electrodeposited photoresist. Conductive surfaces are completely covered with resist by submerging them in an aqueous, micellar dispersion of the resist. The panel is charged to either a negative or a positive potential and attracts the polymer micelles in the resist bath. Lines and spacing at 1 mil (25 μm) have been successfully produced with this technology.

To remove nonreacting monomers and display the base copper, that is, the raw circuit, the exposed panel is passed through a developer. A critical point here is the removal of developer traces from the copper. These are nonconductive and may promote missing plating, step plating, plating voids, separations, and other defects.

1.8
Electroplating

After surface preparation for residual dirty, dry film chips and oxidation removal, the boards are electroplated with copper. The electroplating process produces the plated through holes (PTHs). This step is significant for reliability. The average copper thickness is correlated to specific field applications.

Two approaches can be used to create the final circuit pattern:

- *Panel plating:* In this case, the entire panel is plated to its full thickness before a plating resist is deposited and the etch-resist metal is plated on the board. The subsequent step is to etch away all unwanted copper.
- *Pattern plating:* In this approach, the plating resist is applied before the copper plating and only the final conductor pattern is plated. The subsequent step is to etch away the remaining copper.

There are different copper plating additive systems, all with the possibility to use a direct current or a reverse-pulse current (Dini, 1993). The chemistry of the brightener, of the basic electrolyte recipe, and of the applied parameters allows the control of the copper microstructure, thickness, uniformity, adhesion, and physical properties and of the worst of the defects having this source. Examples of defects are pitting, thickness uniformity, dendrite or macrocrystalline structure, plating cracks, structure modification in roughness form, and so on.

The copper plating may be used alone or as an underlayer for further nickel plating or nickel–gold plating. The nickel–hard gold plating, on a flash of soft gold, is used in applications such as

- Electrical mobile contacts
- Corrosion resistance
- Oxidation resistance in hot and humid environment.

Usual plating is "full gold" or "full-body gold" with thin soft gold, 3–12 μin. ($1 \mu m = 40 \mu in.$), continued with selective hard gold plating, up to 30–50 μin. To plate selectively, a gold mask is applied. This is a dry film similar to the film used for plate etching. Otherwise, it is possible to use a thin layer approximately 100 μin. (2.5 μm) of tin or tin–lead on top of the copper to function as etch resist. Significant in this case is to have a continuous layer of a crystalline structure capable of protecting the circuit during the base copper etching.

To protect the conductive circuit pattern during the process that removes all unwanted copper, in the subtractive type of PCB fabrication, a metal that will not react with the etching agent is plated to it. This, typically, is one of the following:

- Tin–lead, usually a fluoborate solution
- Tin, usually fluoborate or organic sulfonic acid solution
- Nickel–gold using a nickel predip, a nickel bath, gold predip, and a gold plating bath
- Nickel.

Of these, tin–lead is the most popular. However, there is an ecologically rooted movement to eliminate lead from electronic equipment and this has an impact on the development of alternatives to tin–lead solder.

1.9
Copper Etching

The removal of nonuseful base copper starts with film stripping.

Panels entering the etch process have been coated with etch resist, usually a dry film resist. The resist layer selectively protects the circuit area from etch, whereas the remaining copper foil is etched away. The etching agents sprayed onto the surface of the panel remove the exposed copper but cannot dissolve the copper residing under resist. Acid cupric chloride and alkaline ammoniac are the most common etching agents.

The board is passed through a controlled etch rate solution. The chemistry and parameters of this solution are significant to avoid extreme situations, such as underetching, shorts or overetching, reduced lines, and opens. After that, the circuit is cut out, and it is possible to remove the temporary tin protection through a chemical dissolution process, the tin strip in an inhibited acid solution.

1.10
Solder Masking

The solder mask is applied to protect the PCB before hot air solder leveling (HASL) Only a part of the lines and pads have to be soldered during the assembly process. The part of the circuit not utilized for assembly is covered with solder mask. The choice of solder mask, epoxy and half-epoxy or fully aqueous, and its mode of application, screening and spray, depend on the use of the product and also on the possibility of further chemical operation application, that is, electroless nickel immersion gold (ENIG) nickel–hard gold, or other final finishes, where thin layers may be required. The types of circuits, fine lines, flip-chip pads, small dams, and so forth will play an important role for solder mask choice through the imposed undercut and mask definition. Surface preparation before solder mask is important to ensure the good adhesion of the mask by creating an adequate topography. The usual way to do this is by scrubber running a wet abrasive powder or by chemically controlled etching. In all cases, solder mask processing is similar to film processing. This means that there are the same steps of application, exposure, and developing. Due to different chemical composition, the chemistry for developing varies. Basically, solvents or diluted carbonates are used. The exposure of solder mask is critical, and as for all photoimage processes, the alignment is determinant. Final curing of the mask is a key step. In this case, it is necessary to take into account other thermal excursions that will additionally bake the mask, such as nomenclature, carbon ink, or compatibility with further chemical processes. For ENIG, the mask should be undercured for better chemical resistance. Overcured masks are brittle.

1.11
Surface Finishing

Most frequently, the exposed circuit will be soldered. To build or to preserve solderability, the boards will pass though a final finishing step. The HASL, or oil-fused plated solder, has been used for a long time to ensure solderability. As the board design becomes increasingly sophisticated, the solder was found insufficiently efficient as a finish. Other surface finishes appeared, for example, organic solderability protection (OSP), ENIG, immersion tin, immersion silver, electroless or immersion palladium, immersion bismuth, and so on to replace HASL.

Electrolytic plating substitutes for HASL include rhodium, palladium, palladium–nickel alloys, and ruthenium.

To ensure finish application and the reliability of the product, surface cleanliness after the final finish is required.

1.12
Routing

To facilitate the use of boards in assembly, they are routed alone or in groups that are in panel forms, depending on their size. Burr or smear is unacceptable. The boards resulting from this process are of their final size. A postcleaning operation usually with pressurized warm water is used to remove all dust or debris resulting from routing.

There are many other specific steps depending on user demand. Examples are the nomenclature, back print, plug via hole with solder mask, temporary masking, carbon conductive ink applied by screening followed by baking, and so on.

1.13
Testing and Inspection

The final step in board fabrication is to verify the integrity and the functionality of the board. After routing, the PCB is submitted to final control and reliability and quality tests. Control and tests are also interoperational.

The tools for parameter PCB product analysis are the test coupons. These are part of the quality assessment processes that cover reliability evaluations, end-product evaluations, work-in-process evaluations, and process parameter evaluations. There are coupons to test every factor, but the challenge is to test factor interactions in real time.

The main classes of tests are as follows:

- Optical (visual, automatic optical inspection, microscopy under cross-sectioning, scanning electronic microscopy (SEM))
- Electrical (continuity, dielectric withstanding, impedance, and arc resistance)

- Mechanical (tensile, ductility, peel strength, adhesion, bond strength, flexural and impact strength, and dimensional stability)
- Solderability tests (wetting balance, sequential electrochemical reduction analysis (SERA), and dip and look)
- Thermomechanical (glass transition temperature T_g, coefficients of thermal expansion (CTE), Z-expansion, and delamination time)
- Chemical (resistance to chemical, flammability)
- Reliability (thermal shocks, fatigue test, and accelerated stress tests)
- Long-term reliability (surface insulation resistance tests, aging tests, corrosion tests, radiation tests, water absorption tests, and biological tests).

The electrical tests are important since PCBs are electric or electronic devices. The main types of electrical tests and faults are listed in Table 1.1.

The possible mixture of elements to be tested is combinatorial, and so a clear understanding of exactly what is tested and how much is critical to avoid quality and reliability problems during assembly or operation.

1.14
Assembling

The PCB is part of a total electronic circuit packaging system, PCBA. Various types of configurations of components are attached to PCB in assembly. The PCB manufacturer should deliver the product that encounters the assembly requirements.

Performance in automatic assembly is a result of a blend of many ingredients:

- Methodology, that is, board, material, size, and requirements
- Designing the product for automation
- Standards.

Figure 1.1 shows an example of PCB manufacturing process. It contains the main steps in PCB manufacturing in an achievable technology. There are several products resulting from the same flow chart and also several flow charts associated with the same final product.

In principle, the PCB fabrication is analogous to integrated circuit fabrication, to thin layer, nanocircuit, or molecular device fabrication. All these fabrications of

Table 1.1 Electrical tests and fault types.

Test type	Failure type
Continuity	Open
Isolation	Short/leakage
TDR	RF impedance fault
Hi-Pot	Voltage breakdown

TDR, time domain reflectometry; RF, radio frequency.

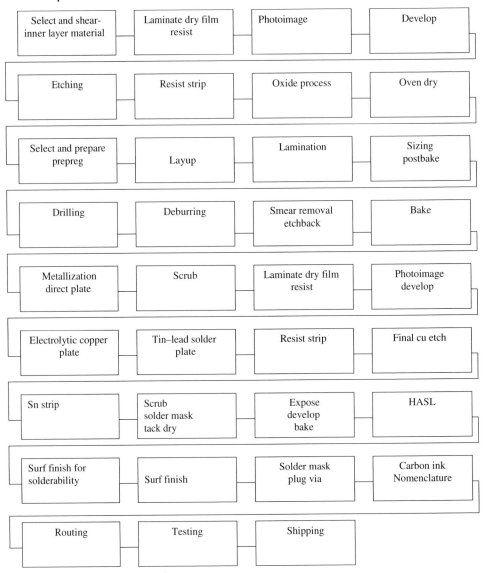

Figure 1.1 PCB manufacturing.

circuits are successions of reversible or irreversible operations of additions and subtractions of new materials joined by masking steps. These operations allow obtaining the designed circuitry.

To outline the similarity in PCB and other circuit fabrication, let us consider in more detail the sequence of manufacturing steps for complementary metal oxide semiconductor (CMOS) process. The starting layup is silicon wafer covered by a layer of oxide and a layer of silicon nitride. On the silicon nitride surface, the photoresist is applied. After masking, the layup is exposed to UV light and to developing. Then the silicon nitride is etched on the nonmasked areas and the photoresist is stripped. The circuitry made by silicon nitride is now visible. For specific areas, a new photoresist is applied. The boron or other atoms are embedded on the nonmasked areas. The photoresist is stripped again. The oxide growing stage is finally performed.

The PCB manufacture starts from a basic layup containing conductive and dielectric layers. This is the result of the former additive process, lamination.

Additive processes include lamination of layers, electroplating, chemical plating, sputtering, screening, implanting, immersion, deposition, adhesion, and so forth. Etching, stripping, dissolution, ablation by laser or plasma, laser drilling, routing, and so on accomplish subtraction processes of existing material. To restrict or enforce the processes of addition or subtraction, to make them selective, intermediate masking steps are required. The available masks are broadly divided into two main categories, permanent and temporary. We should also start considering masks with specified lifetimes. Typical masking in PCB industry is based on resist application followed by exposure and development. Optical light, UV, ion beams, and laser could do the imaging. Tin or tin–lead has the role of masking the copper. Plugs via hole with solder mask and peelable masks accomplish temporary masking. Tapes and floating shields are used in plating processes to mask, completely or partially, specific areas. It is remarkable how many PCB or other circuit constructions are based on a dozen of elementary processes of subtractive or additive type, coupled by well-chosen masking steps.

2
Problem Solving for Reliability and Quality

2.1
Conventional Paradigms

A variety of concepts, paradigms, theories, and methodologies have been applied
successfully for design, quality, and reliability control and, in general, for problem
solving. These include

- Statistical process control (SPC),
- Statistical designs of experiments (DOEs)
- Robust design
- Total quality management (TQM)
- Concurrent engineering
- Quality function deployment (QFD)
- Reliability analysis
- Failure mode effect analysis (FMEA)
- Failure mode effects and critically analysis (FMECA)
- Cause and effect diagrams
- Mind maps, cognitive maps, and conceptual graphs
- Formal concept analysis
- Petri nets
- Case-based reasoning (CBR)
- Theory of inventive problem solving.

SPC is the methodology to detect and quantify quality problems and to keep the
parameters and product in specifications. In SPC, the important controllable
characteristics of a process are determined and their variation is monitored and
controlled by statistical methods. This may result in good quality and reliability of the
product with minimum inspection (Deming, 1982).

Statistical DOE contains the basic methods to organize the experiment to obtain
significant information using a limited number of experimental runs. The science of
statistical DOE originated as a key tool for turning data into information (Box, Hunter
and Hunter, 1978; Montgomery, 1984). Examples of classical designs of experiments
are full factorial, fractional factorial, screening designs, Plackett–Burman, Box–

Evolvable Designs of Experiments: Applications for Circuits. Octavian Iordache
Copyright © 2009 WILEY-VCH Verlag GmbH & Co. KGaA, Weinheim
ISBN: 978-3-527-32424-8

Behnken, central composite designs, orthogonal arrays (Hedayat, Sloane and Stufken, 1999), and optimal designs. Fractional factorial experiments are of interest when not all possible factor setting treatment combinations can be evaluated. Basic tools for DOE are the analysis of variances ANOVA, the regression, and the response surface methodology. ANOVA is the technique for decomposition of variance and for determination of the relative importance of various factors. The regression or correlation shows the strengths of the relationship between input and output of the studied system. Linear or multilinear regressions are more applied. Response surface methodology is a set of techniques designed to find the best value of a response and to map contours of the response surface to gain a better understanding of the overall response system (Lucas, 1994).

Several contacts between the DOE methods and the new artificial intelligence and artificial life methods inspired by the qualitative nature of biologically based information processing have been established. Artificial intelligence typically models the cognitive activities such as problem solving, pattern recognition, or type of thinking. Developed types of DOE are based on tools such as neural networks (NNs) genetic algorithms, expert systems, fuzzy logic, and so forth (Smith, 1993).

Neural networks are information technologies based on the way neurons in the brain collectively process information. They make use of nonlinear mathematics and learn how to model the experiment by using error reduction techniques.

The genetic algorithm is an information technology that imitates the processes of biological evolution with the ideas of natural selection and of survival of the fittest to provide effective solutions for optimization problems. They have been applied to general problem solving, classifier systems, control, and optimization.

Expert systems represent an information technology based on application of rules derived from expert knowledge, which can imitate the intellectual behavior.

Artificial life is a field of study devoted to understanding life by abstracting the fundamental dynamical principles underlying biological phenomena and recreating these dynamics in other physical media such as computers. Artificial life methodology, and how it can impact manufacturing in the future, is the object of several studies (Ueda *et al.*, 2001). Artificial life typically models activities such as self-replication, self-adaptation, and emergence. Agent-based modeling, cellular automata, evolutionary programming, and swarm intelligence represent artificial life specific techniques.

Robust design methods, initiated and promoted by Taguchi (1986), make an innovative use of conventional mathematical formalism of statistical DOE. Robust design objective consists of making product performance insensitive to raw material variation, manufacturing variation, operating environment, and customer's environment.

An important contribution of Taguchi to quality control and improvement is his emphasis on studying not only location effects but also dispersion effects.

Taguchi's statistical approach is rooted in orthogonal designs with modifications that are uniquely his own. Robust design suggests a threefold approach to designing quality into a product or process. The approach starts with the system design, that is, the use of current engineering methods, materials, and processes to build a new system. The parameter design is the identification of key products or parameters and the application of DOE methods to find optimal parameter settings. Finally, the tolerance design that implies the parameter specification limits is performed.

TQM technique starts by using SPC to study variation in products. Identifying the process step causing defects or variability allows the study of the causes of these imperfections. Structured DOE are performed to target the process and identify the parameters that need to be controlled to minimize variation. Then the SPC is of use to monitor the process parameters before communicating the capabilities of the process to designers.

Concurrent engineering and concurrent manufacturing are paradigms that refer to the integrated development of products, the manufacturing processes, and support systems. The basic idea in concurrent engineering is that rather than starting with new product design and then continuing with manufacturing, testing, and application, all the compartments should work together from the beginning. Concurrent engineering must include simultaneous consideration of the domains of marketing, design engineering, manufacturing engineering, production reliability, and quality.

QFD is a structured technique assuring the customer input in any production step. QFD includes planning and deployment matrices, process plans, quality control charts, and operation instructions. This facilitates analysis and provides for an accurate transfer of the customer's requests.

Reliability analysis addresses the changes in quality of a product over time. Reliability may be defined as the probability that an item will perform its intended function for a specified time interval under stated conditions. It is clear from this definition that the reliability level or class has an objective and a subjective basis. Reliability has become as important as quality for manufactured products. Statistical analysis of failure data, life testing, and probability models of survival are important tools in reliability analysis (Condra, 1993).

Several methods utilized to help individuals or groups to describe their ideas in a pictorial form have been identified by names, such as "cause and effects diagrams," "mind maps," "concept maps," " conceptual graphs," cognitive maps," "mental maps," "causal loops," and so forth (Eden, 1988; Novak, 1991). Several of these conventional methods can be represented as hierarchic trees or fishbones, while others need a reticulated network-like presentation.

Conceptual graphs can be considered as a compromise between formal and graphical languages. Due to the graphical representation of knowledge, the conceptual graphs allow for the construction of user interfaces. Conceptual graphs differ from concept maps since the nodes may contain concepts, events, states, goals, and other elements that are described by names or propositions. The nodes are connected by relations. Conceptual graphs are developed with the aid of mapping computer software (Sowa, 2000). Recent developments of conceptual graphs are the flexible modular frameworks that can be tailored to an open-ended variety of architectures for intelligent systems.

Formal concept analysis is a theory of data analysis that identifies conceptual structures among data sets (Ganter and Wille, 1999). A strong feature of formal concept analysis is its capability of producing graphical visualizations of the inherent structures among data. The relationship with conceptual graphs was beneficial for applications in computer science. The Galois lattices that are extensively used in formal concept analysis have been interpreted and applied as classification systems.

Petri nets are mathematical representation of discrete distributed systems. They illustrate graphically the structure of the distributed system as a directed graph with annotations. Petri nets have places, transitions, and directed arcs running between places and transitions. Applications are in data analysis, workflow management, reliability analysis, diagnosis, concurrency, software design, and so forth.

The case-based reasoning is a methodology proposed to problem solving by using previous experiments. Instead of relying solely on common knowledge of a problem domain, CBR is able to utilize the specific knowledge of previously experienced cases. A new experience is retained, and each time a problem is solved, making it immediately available for future problems (Watson, 1997).

The theory of inventive problem solving is a method for innovative approach in problem solving. It quantifies the basic human experience of invention. The theory is supposed to create analytical discipline for inventive problem solving that overcomes basic engineering contradictions step by step to achieve potential breakthrough concepts (Altshuller, 1984).

The methodologies and paradigms presented above are not completely separate. Various tools emphasize on different aspects of overlapping targets, such as design, quality, process improvement, cost reduction, and short cycle time. Some paradigms include main aspects of their predecessors and elaborate on them. Many companies have been encouraged to use these quality methodologies and the related assortment of mathematical techniques. Despite evident successes, the results of the implementation of these methods do not meet the expectations in numerous cases.

Confronted with persistent quality problems, many companies show a trend to lose confidence in predictive or theoretical methods, such as detailed modeling or classical DOE, and make decision mainly by unstructured experience. This may be the source of long-term loss in productivity and competitiveness. To regain industry confidence in theoretical methods, there is a strong need for new methodologies enabled to successfully confront the barrier of complexity.

Main reasons for breakdown of the conventional quality methodology are discussed next.

2.2
Complexity and Time Frames

To clarify the breakdown of many conventional methods for today's industry, it is necessary to take into account the high complexity of the product itself, of the market, and of the resulting change of time frames for both market and production.

There is no univocally acceptable insight into and definition of what a complex system or a complex product is (Kauffman, 1995; Adami, 2002). Several authors reviewed different definitions of the complexity and the associated concept of emergence (Horgan, 1995; Boschetti *et al.*, 2005).

The definition of complexity and emergence seems to change with the domains of application. Some studies consider that complexity has no absolute meaning, and it is only a relative notion dependent on the level of observation. However, it is generally accepted that some products are more complex than others. Despite the fact that most engineering systems are recognized as complex, more of the traditional ones may be operated in a regime where complexity properties can be neglected. The challenge for engineers and entrepreneurs is not only to detect but also to cross the complexity barrier.

Usually, a complex system is defined as a structure or process involving nonlinear interactions among many parts that displays emergent properties. In other words, the aggregate system activity is not derivable from the linear summations of the activity of individual components. Typically, complex systems exhibit hierarchical self-organization in levels under selective constraints.

Features such as nonlinearity, emergence, finitude, hierarchy, timescales, nonequilibrium, unpredictability, interconnectivity, self-organization, collective behavior, evolvability, and multiagency are associated with complexity studies.

Complexity theory has started to play a major role, as modern manufacturing systems and the products ought to be operated as complex systems in the regimes of complexity (McCarthy, Rakotobe-Joel and Frizelle, 2000; Ueda *et al.*, 2001).

Complexity is related to nonlinearity, a necessary but not sufficient condition of chaos, as well as to density, interconnectivity, self-organization, self-similarity, and collective behavior (Mainzer, 1996). The fact that the market is global indicates that the strong interactions and the finitude of the resources are recognized. The new market system is closed, volatile, interconnected, stochastic, and chaotic. The entrepreneurs are no longer willing to wait many years for large experimental studies leading to product change or improvement but seek for rapid market peaks associated with long-term profitability. New methodologies in manufacturing are necessary to produce meaningful results within a definite time span with an imposed budget.

Common to today's manufacturing systems are the problems having a large number of strongly interacting factors. Hundreds of factors are involved in typical problem solving and failure analysis. At the same time, production pressure imposes that large tests with many trials are difficult to be organized and supervised by classical methodologies.

The natural tendency in industry is for a highly integrated product of increasing complexity. Product complexity comes from strong interactions, from emergent behavior, and from the multitude of contradictory requirements.

Following the product and the market complexity, the product design and manufacturing, the quality problems and the product failures became highly complex.

In a typical manufacturing environment, the cause of a component failure is not unique but may be attributed equally to component quality or reliability, to manufacturing process defects, to design marginality, and to the nonlinear interactions between these factors. High complexity means higher costs, more defect opportunities, lower yields, and greater difficulty in fault diagnosis and problem solving.

Unconventional time structures are naturally associated with the complex industry conditions. The multiple timescale issues and the cyclic character of time frames will be briefly illustrated.

Accelerated production and testing methodologies and conventional slower procedures coexist as complementary aspects of industrial complexity. Highly accelerated product design and qualification, the quick turnout, or accelerated failure analysis under emergency conditions is now a typical request for manufacturers.

Modern reliability and quality methods should be part of the design process, quality control must be proactive, and testing must be highly accelerated and tailored to the products' physics of failure. Accelerated testing would not replace but complement the conventional low fatigue reliability testing. Examples of frequently used accelerated tests for Printed circuits boards (PCBs) are

- Accelerated stress tests (ASTs)
- Interconnect stress test (IST)
- Highly accelerated stress test (HAST)
- Highly accelerated thermal shock (HATS).

Accelerated tests are those in which the applied stresses are far above those expected in service but not high enough to induce failures that would never occur in service. DOE utilized in conjunction with accelerated tests as real data source should provide quick and efficient way to determine the quality of a process, or the need for process improvement.

The management of PCB reliability is based on a hierarchy of thermal fatigue tests, performed at different timescales, that is, at different thermal ramps. Predictions and correlation between different tests are permanent challenges.

Time frames should be adapted to the fact that the windows of opportunity for production, for quality improvement, and for failure analysis in today's industry are extremely short. The development of the product design after the invention of a new technology should be highly accelerated. It is necessary to be able to promptly detect production and quality problems, failure mode, and also to implement immediate corrective actions. There is no time to fix reliability and quality issues once the product is on the market. The right information should be offered at the right place and at the right time. Consequently, the cycle times gain increasing significance for continued existence. They are naturally associated with manufacturing and the market.

As a result of executing asynchronously several processes or products without a central control unit, malfunctions in time may emerge periodically in any modern manufacturing. The most common sources of malfunction are capacity limits, feedback loops, and temporal delays. Buffer size represents a common form of capacity limit, leading to nonlinear coupling between devices. Unintentional feedback loops in manufacturing systems such as contamination couplings between chemical processes may affect productivity. Temporal delays in the move of information and materials between processes can lead to high nonlinearity and bring an end to product manufacturing. Clearly, new understanding of time constraints and time frames is necessary to ensure survival in complex manufacturing and market conditions.

2.3
Quasilinearity, Circularity, and Closure

Conventional quality and reliability methodologies make use of branching flow-charts, binary decision hierarchic trees, fishbones, and linear or polynomial correlations, a "quasilinear" methodology. "Quasilinear" character is taken here in the sense of being one-dimensional, since the logical statements are arranged in an ordered, unique sequence resulting in hierarchically structured systems. The mathematical tool to study linear systems is the linear analysis, while that for hierarchical open systems, such as fishbone or hierarchic trees, is the non-Archimedean or ultrametric analysis (Schikhof, 1984; Rammal, Toulouse and Virasoro, 1986). Quasilinear approaches have proven to be inefficient for several high-complexity situations. Quasilinear and ultrametric methodology may provoke global failure in problem solving, despite acting locally in optimal conditions. The conventional cause–effect relationship is quasilinear and for this reason limited. For instance, in the study of phenomena resulting from self-organization among many components, it is pointless to single out only a few factors contributing to a problem to solve. It is hard to distinguish between the cause and the effect in a system with several closed causal loops, nested or interconnected through many dynamical levels, often making the determination of even one single parameter problematic. The cause–effect approach limits can also be seen in situations of atypical failures or anomalies resulting from a succession of accidents. It is the case of failure situations that cannot be duplicated in a testing environment.

Typical quality models, such as cause and effect, FMEA, DOE, neural networks, and ultrametric frames, presented in relation to problem solving and quality control are characterized by

- Causally determined input/output relationship with or without implemented feedback
- Basic linear or hierarchic tree-like, that is, ultrametric organized structure
- Operational openness.

There exists a belief, itself of linear inspiration, that a product and its manufacturability can be improved with the help of standard theoretical tools only by meticulously and extensively implementing them. From the point of view of the modern challenges for "complexity" and "emergence," this belief appears to be questionable. The conventional problem-solving tools implement the quality system as unclosed and therefore nonevolvable and noncognitive (Kaehr and von Goldammer, 1988).

The self-adaptive systems that learn how to learn and that can evolve should be the operationally closed ones (Maturana and Varela, 1979; Albus, 1991). The confrontation of the model with nonlinear reality imposes and at the same time assists this type of closure. In reality, the environment is part of the closed evolvable system.

Concepts of circularity and closure have been developed in the study of biological and cognitive systems. The most important for this book, the semantic or semiotic closure, the functional cycle, and the circular schema concepts, will be briefly presented.

In a significant investigation applicable to both real-life and artificial devices, Pattee pointed out that the evolution requires complementary description of the material and symbolic aspects of events (Pattee, 1995, 2000; Rocha, 2001). Life and evolvability involve a semiotically closed self-referencing organization between symbolic records and dynamics. Symbols, as discrete functional switching states, are seen in all evolvable systems in the form of genetic codes, and at the core of all neural systems in the form of informational mechanisms that switch behavior. Symbolic information such as that contained in genotype has no intrinsic meaning outside the context of an entire symbol system in addition to a material organization that interprets the symbol leading to specific functions such as construction, classification, control, and communication. Self-reference that has evolvability potential is an autonomous closure between the dynamics, that is, physical laws of the material aspects and the constraints, that is, syntactic rules of the symbolic aspects of a physical organization. Pattee refers to this condition as semantic or semiotic closure and concludes that it requires separating and complementing the symbolic description (genotype, design, software, DOE, and logical systems) from the material embodiment (phenotype, machine, computer, real case study, and physical systems). Semiotic closure represents a tentative symbiosis of the logical and physical aspects. The symbolic description must be capable of generating the material embodiment. Symbolic descriptions that generate local dynamics promoting their own stability will survive and therefore serve as seeds of evolvable systems. Finally, the material embodiment must be capable of regenerating cyclically the symbolic description with the possibility of mutation. Cariani (1989, 2001) evaluated the semiotic closure principle relation with the design of devices with emergent semiotic functions. Self-modification and self-construction were recognized as important to the symbol–matter problem and as requirements for semiotically adaptive devices or evolvable ones. In EDOE constructivist approach, the symbolic description is associated with DOE matrices while the material embodiment corresponds to real field measures, data, and products.

The circularity and closure are also related to the *Umwelt* concept introduced by von Uexküll (1928) in theoretical biology to describe how cognitive organisms perceive and interpret their environments. The Umwelt was defined as the part of the environment that an organism selects with its specific sense organs according to its needs (von Uexküll, 1928). Umwelt theory asserts that a complex system does not respond to its environment but rather to its perception of the environment. A complex system actively creates its Umwelt through repeated interactions with the environment. It simultaneously observes the world and changes it, the phenomenon that von Uexküll called a functional circle. The functional cycle includes sensors and actuators. The sensory experience is based on interactions that have specific purposes. The elementary unit of evolvable systems includes the functional cycle of the following four parts: the object, the sensors, the command generator, and the actuator. The Umwelt concept offers suggestions for the study of artificially evolvable circuits.

Circular reactions have also been emphasized in the study of action schema conducted by Piaget (1970). Piaget called his general theoretical framework "genetic epistemology" because he was primarily interested in how knowledge developed in living organisms. Cognitive structures are patterns of physical or mental actions that underlie specific acts of intelligence and correspond to the stages of development. According to Piaget, there are four primary cognitive structures, that is, development stages: sensory motor, preoperations, concrete operations, and formal operations.

The Piaget's action schema, which constitutes the foundation of his learning theory, is a cycle including three elements: a recognized situation, an activity that has been associated with this situation, and an expected result. The recognition of a situation involves assimilation; that is to say, the situation must manifest certain characteristics that the organism has abstracted in the course of prior experience. The recognition then triggers the associated activity. If the expected result does not occur, the organism's equilibrium is disturbed and an accommodation may occur, which may eventually lead to the formation of a new action scheme. The fact that accommodation does not take place unless something unexpected happens relates Piaget's schema theory to the notion of the feedback loop in control theory, to von Uexküll functional cycle, and to semantic closure concepts.

The essentially constructivist perspectives of von Uexküll, Piaget, and Pattee has been utilized here for complex problem solving methodology development. The EDOE associated with a specific problem may not be preprogrammed but is actively constructed and constrained by its own dynamical embodiment and development.

The constructivism is based on the thesis that knowledge cannot be a passive reflection of reality, or a passive application of a formal model, but has to be more of an active and interactive construction. This view has its roots in the philosophy of Vico and Kant. Kant introduced the notion of schema as a collection of actions and thoughts that agents use to interact with the world and to solve problems.

Elements of constructivism may be identified in recent complex problem solving approaches as, for example, Yoshikawa (1989), Drescher (1991), von Glasersfeld (1995), Pierce and Kuipers (1997), Quartz and Sejnowski (1997), Albus and Meystel (2001), Luger, Lewis and Stern (2002), Chaput, Kuipers and Miikkulainen (2003), Riegler (2005), and so forth.

2.4
Advance of Reliability Paradigms

The development of different concepts in reliability and quality domain, paralleling the learning level, is presented in Table 2.1. The learning level characterizes how the products increase their control over environmental issues. Current learning levels incorporate the preceding ones and their associated paradigms. The rise in complexity follows and parallels the increasing learning level.

The first level corresponds to basic reliability control. This is statistically oriented. The manufacture produces high volume and monitor how many products come to control with reliability problems. The control may be internal or external product screening in field conditions. From that information, if the reject percentage is considered as unacceptable, the higher level of learning, that of reliability analysis, is

Table 2.1 Reliability paradigms evolution.

Concept/year	Before 1970	1970–1980	1980–1990	1990–2000	2000–2010	2010-Future
Learning and control level	Low learning	Supervised learning	Unsupervised learning	Learning to learn, self	Proactiveness, predictive	Evolvability, intelligence
Approach, culture	Product screening	Conformance tests	Reactive feed-back	Reactive self-closed	Proactive feed-before	Innovative, creative
Methods, source	Practice	Matter sciences	Life sciences, biology	Cognitive sciences	Mathematics, logic	Living systems
Reliability paradigm	RC	RA	Reliability MS, RMS	Self-adaptive RMS, SARMS	Proactive RMS, PARMS	Evolvable RMS, ERMS
Goal for reliability	Output	Max. output at min. cost	Metrics or procedures	Self-modify metrics	Proactive procedures	Evolvable procedures
Reliability, driving force	Functional product	Assurance or specification	Assurance, low cost	Customer satisfaction	Customer needs	Evolvable environment
Problem solving	Irregular	Problem focus, fix	Corrective actions	Self-corrective	Proactive action plans	Evolvable action plans
Manufacture	Craft manufacture	Mass manufacture	Lean manufacture, just in time	Agile manufacture	Customized proactive	Evolvable manufacture
Circuits	Simple fixed	Conventional circuits	Adapted circuits	Self-adaptive circuits	Proactive circuits	Evolvable circuits

R, reliability; MS, management system; PA, proactive; SA, self-adaptive; E, evolvable; RC, reliability control; RA, reliability analysis.

implemented. Several internal tests are performed before shipping the product and where the root cause is addressed. At this level, physics of failure is the investigation tool.

The next level of learning was that of management of reliability. The focus is the product life and the approach is still reactive. Biological sciences represent a source of inspiration for the associated methodologies. The implementation of standards is significant at this level. Reliability management system (RMS) establishes the failure mechanism and the corrective actions for known nonconformities. Repairing methods are naturally covered here.

The next level of learning includes the self or auto concepts, that is, the implementation of a closed or cognitive-like system capable of managing the reliability without external contributions. Self-adaptive reliability management system (SARMS) is the informational closed, relatively autonomous variant of RMS. Any failed product has an already implemented correction. Cognitive sciences represent the main source of associated methodologies. At this level, the learning to learn is operational. The reliability system is designed to be self-adaptive. Self-controlling and self-repairing methods are included here. These open the road to evolvability.

The previously presented learning levels and the associated concepts are based mainly on feedback and reactivity.

This corresponds to the situation in which the industrial product is manufactured and then gradually improved taking into account the resulting quality and reliability. A new level in reliability is that in which before manufacturing of any new product, a proactive reliability system is implemented to prevent possible nonconformity of an unforeseen type, that is, virtual failures.

The proactive reliability management system (PARMS) is based on mathematical and logical models.

The future level in the development of the reliability concepts may be that of evolvable reliability models, embedded and operated intrinsically. This level should go decisively beyond the learning, in the domain of evolvability, creativity, and intelligence.

The manufacturing system has a parallel development from craft production, to mass production, to lean production, and to agile production. Possible next steps will correspond to PARMS and then to evolvable reliability management systems (ERMSs).

Different types of circuits correspond to different learning or postlearning levels. As the level of learning and control increases, the circuit has an evolution from fixed circuits toward conventional circuits, adaptive circuits, self-adaptive circuits (SACs), proactive circuits (PACs), and evolvable circuits (ECs).

Part Two
Evolvable Designs of Experiments (EDOE)

Evolvable Designs of Experiments: Applications for Circuits. Octavian Iordache
Copyright © 2009 WILEY-VCH Verlag GmbH & Co. KGaA, Weinheim
ISBN: 978-3-527-32424-8

3
Polystochastic Models

3.1
What Is PSM?

The term polystochastic was coined to describe stochastic systems that emerge when several stochastic processes are enabled to interplay with each other, resulting in a system that is structurally different from its components. The first objective of polystochastic models (PSMs) was the study of the structure of the stochastic evolution on more conditioning levels. Basic frames have been presented in the research monograph from 1987 (Iordache, 1987). The initially exploited frames, restricted to the real field, were later generalized to outline the interaction between real field and "other than real" algebraic fields in PSM (Iordache, 1992). Starting from 1990, PSM was applied as theoretical tool for problem solving in the domain of industrial engineering systems. PSM have been tested for diagnosing, classification, failure analysis, flow-sheet generation, and for technology and engineering design improvement.

The main mathematical tools used for PSM were the stochastic systems as, for instance, "random evolutions" (Keepler, 1998; Hersh, 2003) "random dynamical systems," "random systems with complete connections (RSCCs)" (Iosifescu and Grigorescu, 1990), and "random iterated function systems" RIFS (Barnsley, 1993). Finally, PSM have been presented in the unifying frame of category theory (MacLane, 1971).

The theory of random evolutions has as objective the study of a significant class of stochastic dynamical systems. In this theory, random means not only stochastic inputs or initial conditions, but also random media and stochastic process in the equation of state. PSM makes use of random evolutions to describe phenomena in which several component stochastic processes are connected by the so-called control chain describing the evolution of the environment that induces the switching from a component process to another. Random evolutions describe situation in which a process controls the development of other processes, the other processes being described as operators. This is the situation considered by the PSM in which the control process of conditions connects the component stochastic process associated with the operators. The discrete control process of conditions determines the switching from a component process to another. Connections between random

Evolvable Designs of Experiments: Applications for Circuits. Octavian Iordache
Copyright © 2009 WILEY-VCH Verlag GmbH & Co. KGaA, Weinheim
ISBN: 978-3-527-32424-8

evolutions, products of random matrices, and random processes in random environments were outlined by Cohen (1979).

The following PSM presentation keeps the structure and the notations as close as possible to that used for general learning model or RSCC but it should be emphasized from the start that the significance of the PSM elements and the calculus is different. The general learning model or RSCC aims to describe processes in which the subject is repeatedly exposed to an experimental situation that is a condition or an event, for which various responses are possible, and each trial can modify the subject's response tendencies. The set of all possible states is denoted by s, the set of conditions by k. Let k_n be the condition in the trial n when the subject state is s_n. Let $p(k_n|s_n)$ be the probability of the condition k_n, conditional on the state s_n. The state at the $n + 1$ trial will be $s_{n+1} = u(k_n, s_n)$. The operator $u: k \times s \rightarrow s$ characterizes the system tendency. The elements of the general learning model are quadruples $\{s, k, u, p\}$. This basic model has been comprehensively studied in automata theory, control theory, learning theory, and mathematical theory of RSCC (Iosifescu and Grigorescu, 1990).

Figure 3.1 illustrates the general learning model or equivalently the RSCC frame.

The initial system state is s_0. Denote by $p(k_0|s_0)$ is the probability of the condition k_0, conditional on the state s_0. The resulting state at the first trial will be $s_1 = u(k_0, s_0)$, and so on. The differences between such well-established stochastic systems as RSCC and the PSM are imposed by the fact that PSM objectives are different. PSM intend to study stochastic systems that go beyond learning and adaptation, more specifically to evolvable or cognitive systems. PSM intends to be a liable framework for complexity and evolvability studies.

Some aspects illustrating the specificity and originality claims of the PSM approach are emphasized in the following: The first aspect is the vectorial character. This is related to system hierarchical organization in several conditioning levels and scales. This organization is the strategy to confront complexity in several systems, technological, biological, and cognitive. Most real-world complex systems have both temporal and spatial structure and are hierarchical in nature. The focused elements of PSM will be quadruple of vectors $\{S, K, U, P\}$ referred here by the term "$SKUP$." The notations are S-states, K-conditions, U-operators, and P-possibilities.

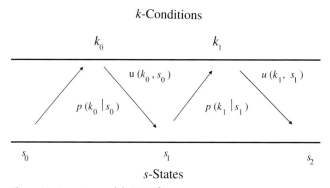

Figure 3.1 Learning model, RSCC frame.

The *SKUP* elements are vector valued. This enables to describe sequential and parallel evolutions. One level that is one-scale, sequential systems may characterize systems that learn, but are not capable to describe novelty and creative evolution.

The states S, the conditions K, the operators U, and the possibilities P will be vectors denoted as follows: $S = (s^0, s^1, \ldots, s^m, \ldots, s^M)$; $K = (k^0, k^1, \ldots, k^m, \ldots, k^M)$; $U = (u^0, u^1, \ldots, u^m, \ldots, u^M)$; $P = (p^0, p^1, \ldots, p^m, \ldots, p^M)$.

Here, s^m represents the particular state at the level m, and k^m represents the particular condition at the conditioning level $m \leq M$. Typically, upper indices are reserved to levels, while lower indices are reserved to time steps. The components of U may be operators such as $u^m: k^m \times s^{m'} \to s^{m''}$ or $u^m: k^m \to s^{m'}$, for example. The operator U should be able to accommodate change of conditioning level and the splitting of these levels.

The next innovative aspect of PSM is a complementarity issue. S and K may be defined on different types of algebraic frames. Frequently, S is defined on the real field R, whereas K is defined on "other than real" algebraic frames, for instance, over finite fields, local fields, or algebras. Such complementary frames are crucial to ensure semantic closure and evolvability. Observe that despite algebraic frame differences, S and K should be interconnected. This interconnection is described by operators U and possibilities P. U characterizes the K to S transition and P characterizes the S to K transitions, that is, $U: K \to S$ and $P: S \to K$. The role of complementarity for evolvability of living and cognitive systems and for modeling relations was highlighted by Pattee (1995, 2000) and Rosen (1991).

The third innovative aspect concerns the possibilities P. It consists in complementing or replacing probabilities p, by possibilities as, for example, those defined in fuzzy logic (Klir and Yuan, 1995; Dubois and Prade, 2001). Conventional probabilities are inappropriate to illustrate qualitative concepts as plausibility, beliefs, anticipation, guessing, partial truth, and opportunities, all having significant role in complex problem solving. Another reason of this replacement is the need for simple calculations.

The possibilities P are defined by vectors such as $P(K) = (p(k^0), p(k^1), p(k^m) \ldots p(k^M))$. P should be defined as simply as possible. The component $p(k^m)$ is an evaluation of the potentiality of the condition k^m. An example of simple possibility is to take $p(k^m)$ equal to 0 or 1. The value "0" corresponds to situation in which that condition k^m is inactive while the value "1" corresponds to active conditions. $P(K)$ is associated with a binary string in this case. An example of associated vector is $P(K) = (011\,000\ldots)$, where $K = (k^0, k^1, \ldots, k^m, \ldots, k^M)$. This means, for instance, that only the second level, k^1 and the third level, k^2 are activated eventually with probabilities: 0.25 and 0.125 corresponding to the rank. It is a real valuation for condition K keeping some basic properties of probabilities. An ultrametric valuation $P(K) = 2^{-m}$, where m is the index of the first nonnull component in K may appreciably simplify calculus (van Rooij, 1978). $P(K) = 0.25$ for the above example.

The fourth innovative aspect concerns the differential model for K process. The elements of K are resulting as solutions of a partial differential equation. The model proposed here is a wave equation, WE, presented in Chapter 4. Its solutions help in describing the space of conditions K, in establishing the actual form of operators U, ensuring the highly economic character of transfer of copied information from one step to another for evolvable systems.

Significant innovative aspects are related to the PSM presentation in the category theory frame.

3.2
Basic Notions for Categorical Frame

The learning models outline a set of states s, a set of conditions k, and transitions relations between them, expressed by operators as u, and probabilities p.

Figure 3.1 shows the two chains of random variables s and k and their interconnection by u and p. The categorical approach for PSM appears as a natural generalization of such transition systems.

Category theory is the conceptual framework suitable for algebraic structures comparison and classification, for study of interactions (MacLane, 1971; Barr and Wells, 1995).

Category theory provides a common language and a set of unifying concepts to various branches of sciences and engineering. It is a request to make use of category theory, as mathematical frame for a future engineering science that will help produce high quality and reliable products, to ensure durable development.

It should be noted that the categorical approach presented here is deliberately intuitive and informal. Basically, a category contains objects and morphisms associated with arrows such that there exists an arrow from each object to itself, the arrows may be composed and the arrow composition is associative. Examples of objects are structures, processes, sets, concepts, and so forth. The category of sets, denoted by Set, has sets as objects and functions between sets as arrows.

Consider the category in which the objects are categories and the morphisms are mappings between categories. Mappings between categories preserving the categorical structures, namely, identities and composition, are called functors.

A category can be seen as a diagram, that is, a graph, where objects are the vertices of the graph and morphisms are the paths in the graphs. A diagram commutes, if for all paths with equal domain and codomain, the value of the diagram functors is equal. This expresses the fact that the results of the compositions are equal. Commutative diagrams represent the categorical equivalent of a system of equations, but are more general in nature. Significant for the models studied here is not only the commutativity but also the cyclical or closed character of diagrams. The diagrammatic presentations specific to category theory may be adapted to take into account the representations used by engineers as designs, flow charts, and graphs.

A natural transformation is a morphism between two functors. It provides a way to switch from one mapping of a structure to another in a way that is interchangeable with the two images of any morphism. The naturality allows holding functorial implementation together and the knowledge coherence across multiple regions, scales, and hierarchies.

The strength of category theory lies in the possibility of universal constructions as limits, colimits, and exponents. The colimit is a formalization of assembly of objects and morphisms. A colimit for a diagram can be thought of as a structure that completes the diagram to a minimal commutative diagram containing it. The colimit puts everything together. It describes gluing, fusion, and amalgamation. In the category Set, the colimit corresponds to the least set. Limits are the dual notion to colimits, which is the one notion obtained from the other by reversing the arrows and interchanging the initial and terminal for objects. Intuitively, a limit extracts the abstraction. Given a diagram, an element is called a limit if there are morphisms from that element to all vertices of the diagram, and if for any other element satisfying the same property there is a unique morphism from it to the limit. In the category Set, the limit corresponds to the biggest set.

Others universal constructions are the exponents. Exponentiation is a generalization of functions set in conventional set theory.

The coproduct and the product represent the categorical notions corresponding to disjoint union and to Cartesian product in the category Set. The coproduct is a special type of colimit and the product is a special type of limit. The pushout gives composition of objects having the same domain under two morphisms. The pullback gives decomposition of objects having the same image or codomain under two morphisms.

A Cartesian closed category is one that is closed under all kinds of universal constructions, for example, limits, colimits, and exponents.

To any canonical construction from one type of structures to another, an adjunction between the associated categories, will correspond. Adjoint functors are pairs of functors that stand in a particular relationship with each another. A functor can be left or right adjoint to another functor that maps in the opposite direction. A pair of adjoint functors typically arises from a construction defined by a universal property; and it can be seen as a more abstract and powerful view on universal properties.

If F and G represent a pair of adjoint functors, with F left adjoint to G right adjoint, then the composition GF will be a monad. The categorical dual of monads, FG will be a comonad. Every adjunction gives rise to a monad.

A monad or triple is a functor from a category to itself, in other words, an endofunctor. In general, the adjunctions relate categories of different natures. The monad theory tries to capture what that adjunction preserves.

The monads generalize closure operators on partially ordered sets to arbitrary categories.

A monad that was investigated by Giry (1981) as part of the categorical foundation of probability theory is of interest here due to its close connections to PSM. The monoidal category studied by Joyal, Street and Verity (1996) provides a category for abstractly capturing feedback, recursion, and cyclic structures. It is expected that such type of categories will be useful for PSM studies since the monoidal category elements are straightforwardly related to the elements of the quadruple, SKUP. The SKUP associated with PSM outlines the very general architecture shared by the operational structure of evolutionary devices, the functional organization of organisms as informational and cognitive systems, and the scientific and engineering

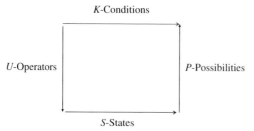

Figure 3.2 Categorical frame for *SKUP*.

method (Cariani, 1989). The diagram from Figure 3.2 shows the *SKUP* that contains two categories *S* and *K*, and two transforming functions, that is, two functors, *U* and *P* between these categories.

The conditions *K* represent the category describing the component process types in PSM. In this case, the process types are the objects. Interactions among the types can be modeled as morphisms. Constructions in category theory such as colimit characterize fusion. *S* is the category, indexing the detailed states of the component processes. The arrows, that is, the morphisms describe the transition relations in component processes. Disjoint algebraic frames for states-*S* (real, dynamical, analogical, and natural) and conditions-*K* ("other than real," symbolic, digital, and formal) have to be considered.

Operators *U* are accounting for interactions in *K*, and between *K* and *S*.

Possibilities *P*, supplement the probabilities to express potentiality, fuzziness as uncertainty, opportunities, and emergence.

Figure 3.2 is an adjunctive diagram illustrating the elements of the *SKUP*.

The two categories, *K* and *S* are interconnected by the functors *U* and *P*.

The symbolic process, in *K*, is said to connect with the dynamical process in *S*, if given the functors *U* and *P*, the diagram from Figure 3.2 commutes.

The arrows orientation for *K* and *S* depends on the application context. The category *K* with the left adjoint functor *U* and the right adjoint functor *P* defines the monad *PU* and the comonad *UP*.

Monoidal categories associated with PSM are effective to describe the cyclical that is the closed character of studied systems. The elements *K*, *U*, and *P* are associated with the monad or triple, while *S* is associated with the so-called trace.

3.3
Illustrative Examples of PSM and Categorical Frames

3.3.1
Lumped Stochastic Chains

This example is related to the study of lumpability for stochastic chains associated with compartmental models (Kemeny and Snell, 1960; Iordache and Corbu, 1987). In such models, the category *K* corresponds to the transition matrices for the classes or the lumps, while the category S corresponds to the transition matrices for the elementary states of the component stochastic processes. Observe that the evolution in K is on a slower timescale than that in S.

To prove that S and K are categories, it is necessary to show that the identity, the associativity, and commutativity of the morphisms hold in S and K. These properties are obvious for the stochastic matrices, that is, in S. In K, the selected lumping of states method should ensure similar properties for morphisms.

The challenge is to define functors that will be a natural way to transform the category K of lumped matrices into the category S of stochastic matrices and also functors that will allow to transform S back to K. The process in K models the transitions between the states in K associated with the transition processes in S, if the defined functors U and P ensure the commutativity of the diagram from Figure 3.2.

For the studied lumping models, the functors P and U are product of matrices. To give an example, let us denote the stochastic matrix in S by R and the lumped matrix in K by $\hat{R}$. The definition of P is, $P: R \rightarrow \hat{R}$ with $\hat{R} = BRC$. Here, B is a distribute matrix such that its kth row is a probability vector whose no null components describe the repartition of the initial states in the lumped chain. C is a collect matrix such that the no null components of its kth columns are equal to "1" and corresponds to the states in the initial chain (Kemeny and Snell, 1960; Iordache and Corbu, 1987). The Markovian character of the S chains does not ensure the Markovian character of the K chain of lumped states, except in very special cases, that is, if $CBRC = \hat{R} C$.

To prove that P and U are functors, it is necessary to show that they preserve identity morphisms, the composition of morphisms and commutativity between the categories.

3.3.2
Conditional Stochastic Chains

Another illustration of the categorical frame for PSM is the study of conditional stochastic chains. PSM started by considering systems to be organized hierarchically in levels as arrays of systems within systems (Iordache, 1987). As for the other approaches to complexity, it was assumed that the complexity could be managed through hierarchical layering. The conditioning levels are correlated to time and space scales.

For simplicity of presentation, we make use here of Markov stochastic processes, with one step memory, as the basic suggestion for the construction of PSM.

The first level, $m = 0$, is considered as nonconditioned. Suppose that the process described on K has k states. This is the so-called control process. Let C be the corresponding transition matrix. There are k different choices for the component processes at the next level, $m = 1$. A "component" process defined on $S_1, S_2 \ldots S_k$ corresponds to each state for the center process described by C. The process on S_i is described by D_i, an $s_i \times s_i$ stochastic matrix. The evolution at the level $m = 1$ depends on the evolution on the level $m = 0$. This illustrates the term of the conditional level and that of the conditional process. The process defined on K appears as the process of conditions that switches the evolution from a component process to another. This switching may be induced by external or internal changes. The state in component processes evolves in time and on the other hand, their hierarchical scheme evolves too. The evolution at $m = 0$ is on a slower timescale than that at $m = 1$.

The PSM structure is portrayed by indicating the conditioning levels, $0, 1 \ldots m$ and the space of states to each level, for instance, the states $(1, \ldots, k)$ at the first level $m = 0$, the states $(1, \ldots, s_1, 1, \ldots, s_2, 1, \ldots, s_k)$ for different component processes at $m = 1$. Suppose $s_1 = s_2 = \ldots = s_k = \ldots = s$. Let D denote the common transition matrix of the component processes. The transition matrix C^*D with "*" being a tensor product describes the entire process at the level $m = 1$. It is a stochastic matrix too. The resulting processes have been studied as conditional stochastic processes.

Observe that the tensor product "*" connects different conditioning levels. The same method of construction, via the tensor product "*", may be continued for a number of levels.

Distinctive PSMs arise if K- and S-processes are vectorial and are defined on different types of algebraic frames, that is, on different categories. This implies parallelism, multiple scales, interactions, and closure. The operators U characterize the conditioned transitions for different component process, $S_1, S_2, \ldots, S_k$. They help to detect interactions between K and S. The interactions are critical for PSM definition and applications.

Mechanisms to characterize emergence and to limit the number of states in K as in S are mandatory in PSM study. Significant techniques are based on the tensor product, "*" interpretation as categorical product or as coproduct (Joyal, Street and Verity, 1996).

Figure 3.3 shows the adjunctive square diagram containing the elements of the $SKUP$ and outlining the two possible ways in K, the categorical product "×" way and the coproduct "∪" way. This appears as a second structuring for the category K, mirroring the primary $SKUP$ structure. The switch from product "×" to the coproduct "∪" and the reverse is possible but the two types of categorical product cannot be observed simultaneously. This switching proves to be a significant part of the evolvability and emergence mechanisms.

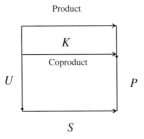

Figure 3.3 Categorical frame for $SKUP$-product and coproduct.

4
First-Order Wave Equation

4.1
Algebraic Frames for Time "T" and Space "Z"

The classification and pattern recognition are the key ingredients in data processing and problem solving for both natural and artificial evolvable systems. Biosystems and cognitive systems do not survive in environments that they do not recognize or misclassify.

Preliminary attempts for pattern recognition modeling by differential equations outlined the significant role of orthogonal arrays (Iordache, 1992, 1996). An intriguing result was that the pattern recognition methods parallel screening procedures in experiment design and in problem solving. In particular cases, one obtained as solutions of the first-order wave equation (WE) orthogonal array matrices, Walsh matrices W, or Latin squares L. Models of cognitive processes such as pattern recognition prove to have as solutions logical thinking methods that are applied in designs of experiments (DOEs). The result emphasizes the profound relation between evolvability and cognition as presented in a constructivist perspective. Both evolution and cognition use the same set of techniques.

To establish the analogous of a dynamical model for classification or pattern recognition, the concept of time and space, in the space of conditions K, will be revisited and adapted to the objectives of this study.

The point of view adopted here is that the significance of the process parameters should agree first with the mechanism and then the nature of analysis of the system studied. Less conventional mathematical frameworks are acceptable if these frames can naturally describe the system evolution, and the system analysis can proceed on this basis.

Evolvable and cognitive systems study needs appropriate concepts for time and space. In physics and chemistry, the real field R plays a dominant role. However there are many other algebraic fields and rings that have a different type of valuations (van Rooij, 1978). The finite frames such as Galois fields and rings or cyclic groups, with trivial valuation, represent the appropriate mathematical frame to describe the finite, logical, or cyclic types of evolution and cognition processes.

Timescales and time cycles should be taken into account. A different understanding of the time frame, a cyclic frame complementing the usual linear one, proves to be

Evolvable Designs of Experiments: Applications for Circuits. Octavian Iordache
Copyright © 2009 WILEY-VCH Verlag GmbH & Co. KGaA, Weinheim
ISBN: 978-3-527-32424-8

necessary. Evolvability requires both the quiescent "timeless" or "cyclical" K-processes and the relatively fast "dynamical" or "linear" S-processes (Pattee, 1995). Finite fields represent a common choice for K, whereas the real field is the commonplace structure for S. There is a natural hierarchical or cyclic structure associated with finite fields, and this explains, in part, why they are considered as the appropriate tool for systems structured in conditional levels.

The first-order wave equation describing the evolvable designs of experiment would contain parameter analogues to the space and time from the standard dynamical mathematical models in physics and chemistry.

Consider, for example, the space Z of vectors describing the properties of an object to be classified and the time T describing the degree of advancement of the pattern recognition, classification, or development for that object. For the classification process, it is possible to associate with different steps in a classification scheme digits such as "0" or "1", with significance as "no" or "yes", "true" or "false," "separated" or "nonseparated", "recognized" or "nonrecognized", and "developed" or "nondeveloped" (Iordache *et al.*, 1993a, 1993b; Iordache, Corriou and Tondeur, 1993).

For any object to be classified, a vector Z will be associated in which the properties are specified by digits in the hierarchical order of significance for classification. The same vector Z will give a description of the classification stages in the associated pattern recognition scheme. Z describes pattern recognition or stages in problem solving or development stages for organisms and so on. Denote $Z = z_0 z_1 \ldots z_j$.

The component z_j should specify the presence of an attribute in classification step, its absence, and also partial or uncertain results.

This means that the mathematical frame cannot be limited to that of dyadic, that is, to Boolean calculus. The need for multivalued characterization of classification steps and objects, the uncertainty, requires mathematical tools to complete the better studied dyadic calculations. Detailed description of dynamical systems need vector characterization corresponding to multivalued logic such as "1", "2", "3", and so on, meaning for instance, low, average, high, and so on. The coordinate "z_j" characterizes the properties, and it is naturally associated with a stage of classification schemes that use the difference in properties noted by "z_j" to perform that kind of classification, pattern recognition, or development.

The degree of advancement in the classification, in the pattern recognition, or in the development was defined as the necessary level of similarity T between two objects represented to be classified in the same class (Iordache *et al.*, 1993a; Iordache, Corriou and Tondeur, 1993). It may be an expansion of the type: $T = t_0 t_1 \ldots t_j$ with digits $t_j = 0, 1, 2$, and so on. Denote also this vector by $T = (t_j)$. Each value of T corresponds to another potential step in pattern recognition or in development. Single-component vectors with modulo-m algebra structure will be discussed as an example. This is one of the weakest algebraic structures for T and Z still providing a mathematically tractable model adequate for classification and pattern recognition operations or for development study. A slightly different frame to be considered is that of Galois finite fields. Recall that finite fields with the same number of elements are isomorphic.

Table 4.1 Sum and product in $C(m)$.

$C(2)$

$(x+y)\,\mathrm{mod}2$				$(x\cdot y)\,\mathrm{mod}2$		
$\oplus$	0	1	$\otimes$	0	1
0	0	1	0	0	0
1	1	0	1	0	1

$C(3)$

$(x+y)\,\mathrm{mod}3$					$(x\cdot y)\,\mathrm{mod}3$			
$\oplus$	0	1	2	$\otimes$	0	1	2
0	0	1	2	0	0	0	0
1	1	2	0	1	0	1	2
2	2	0	1	2	0	2	1

$C(4)$

$(x+y)\mathrm{mod}4$						$(x\cdot y)\,\mathrm{mod}4$				
$\oplus$	0	1	2	3	$\otimes$	0	1	2	3
0	0	1	2	3	0	0	0	0	0
1	1	2	3	0	1	0	1	2	3
2	2	3	0	1	2	0	2	0	2
3	3	0	1	2	3	0	3	2	1

Examples of addition and product tables are presented in Tables 4.1 and 4.2. Here, "$\oplus$" denotes the addition and "$\otimes$" denotes the product. The sum and product refer to component-wise operations for vectors Z or T in K. $C(m)$ denotes the modulo-m algebraic frame and $GF(m)$ denotes the Galois field of order m. $C(m)$ enables to fit the physical intuition concerning the cyclic character of the classification operations in m steps and to justify this first choice for algebraic frame. If $m=2$, the sum $\oplus$ is defined as follows: for any two elements $T=(t_j)$ and $S=(s_j)$, the dyadic sum is $(t\oplus s)_j=(t_j+s_j)\,\mathrm{mod}2)$. This means that $1\oplus1=0$, $1\oplus0=1$. The sum is the dyadic addition $\oplus$, equivalent to the dyadic difference. The rule of addition $\oplus$ signifies that two identical digits have no effect on the classification. Only the difference in digits makes a contribution. This addition appears rather as a comparison than as a sum. The product $\otimes$ is introduced in a way related to cyclic operations too. Product definition takes into account that after m steps the classification process restarts. For the "time" T or "space" Z, no change should happen after completion of a cycle of classification operations.

Another algebraic frame to be considered is the finite field, $GF(m)$. If m is not a prime number, we are faced with rings instead of fields. This algebraic frame was

Table 4.2 Sum and product in GF(m).

GF(2)

$(x \oplus y)$				$(x \otimes y)$		
$\oplus$	0	1		$\otimes$	0	1
0	0	1		0	0	0
1	1	0		1	0	1

GF(3)

$(x \oplus y)$					$(x \otimes y)$			
$\oplus$	0	1	2		$\otimes$	0	1	2
0	0	1	2		0	0	0	0
1	1	2	0		1	0	1	2
2	2	0	1		2	0	2	1

GF(4)

$(x \oplus y)$						$(x \otimes y)$			
$\oplus$	0	1	2	3		$\otimes$	0	1	2	3
0	0	1	2	3		0	0	0	0	0
1	1	0	3	2		1	0	1	2	3
2	2	3	0	1		2	0	2	3	1
3	3	2	1	0		3	0	3	1	2

extensively applied in formal logics. The operations in GF(3) and GF(4) are presented for illustration purposes in Table 4.2.

Let Y denote the range of the output of a system that performs classification based on the examination of feature properties. Y is element of the same algebraic frames as T or Z, may be single-dimensional vector and may assume values 0, 1, 2, and so on, corresponding to various outputs. Multidimensional values such as $Y = y_0 y_1 y_2 \ldots y_j$ should be examined too. Y, like T or Z, is represented by finite strings. The definition of Y needs to ensure the logical consistency of the frame. Appropriate algebraic structures for the range of Y are the fields of real numbers, the modulo-m algebras, or the finite Galois field GF(m) that provides physically significant and mathematically tractable models. The dyadic differential calculus was initially developed for GF(2) situations (Harmuth, 1977). If m is prime p, the range Y is the standard frame for multivalued logic. Single-dimensional vectors T, Z, and Y are useful if the classification process is based on a single property. For multilevel situations, T, Z, and Y would be tensor products "*" of the individual level cyclic groups (Bochmann and Posthoff, 1981; Yanuschkevich, 1998).

4.2
The First-Order Wave Equation

The simplest mechanism of pattern recognition, classification, or development is the one in which "small" changes of the degree of pattern recognition ∂T are associated with "small" changes of the answer ∂Y. It should be emphasized that the differential is in fact a difference since T and Y are discrete.

Moreover, the change of answer ∂Y depends on both the existing answer Y and the change ∂T of T, that is,

$$\frac{\partial Y}{\partial T} \propto Q(Y) \tag{4.1}$$

It is supposed that ∂T is nonnull. Otherwise, the differential equations are replaced by difference equations. The rate of pattern recognition or classification is denoted by Q. This mechanism is of "kinetic" type.

Another classification mechanism takes into account that the variation of the answer Y, along with the degree of recognition T, is proportional to the answer variation along the features space Z. Classification, pattern recognition, and development mean in fact travel in time T along the space of properties Z. As Z is screened with a velocity V, the degree of pattern recognition varies proportionally.

This means that

$$\frac{\partial Y}{\partial T} \propto V \otimes \frac{\partial Y}{\partial Z} \tag{4.2}$$

Here, the velocity is a vector $V = v_0 v_1 v_2 \ldots v_j$ or $V = (v_j)$. This mechanism is of "convection" or "drift" type.

The general model of the pattern recognition process including both types of recognition processes, corresponding to the evolution according to T and Z, is the first-order wave equation:

$$\frac{\partial Y}{\partial T} \oplus V \otimes \frac{\partial Y}{\partial Z} \oplus Q(Y) = 0 \tag{4.3}$$

The initial condition is

$$Y(Z, 0) = F(Z) \tag{4.4}$$

Obviously, V and Q may depend on T and Z.

The fact that the addition is equivalent to the difference suggests that a second-order wave equation does not give new solutions in K, as defined.

The obtained first-order wave equation is formally similar to the simple wave propagation model applied in different domains of physics or chemistry; for example, in fluid dynamics and transport phenomena.

The mathematical formalism for modeling conditions K apparently follows that of the real states S as a first-order wave equation but with different addition and product operations taking into account the corresponding domain. Symbolic models in K and dynamical models in S are complementary in the sense that neither type of the model is reducible to the other. Both are necessary for understanding the whole system including classification schemes and dynamic processes.

Particular solutions of the general model (Equations 4.3 and 4.4) will be presented in what follows.

4.3
"Kinetic" Model: Walsh–Hadamard Matrices

For $V = 0$, the first-order WE reduces to

$$\frac{\partial Y}{\partial T} \oplus Q(Y) = 0 \tag{4.5}$$

The solution in GF(2) is presented here for illustration purposes. In GF(2), "0", denotes the null element. The real product and the sum were translated toward GF(2) operations.

Suppose that the rate of pattern recognition Q is the constant expansion denoted by

$$Q = q_0 q_1 q_2 \cdots q_j \text{ or } Q = (q_j)$$

The solution similar to Euler solution for differential equations will be

$$Y(T) = Y(0) \oplus Q \oplus T \tag{4.6}$$

Recall that the sum $\oplus$ is equivalent to the dyadic difference.

Suppose that $Y(0) = 1$. In this case, the solution of the first-order WE for different Q is $Y(T, Q)$ as shown in Table 4.3. Denote the resulting "0" by "-1" with the same logical signification, for instance "no". Table 4.4 replaces Table 4.3.

Suppose that Y, T, Q are vectors with two components: $Y = y_0 y_1$, $T = t_0 t_1$, $Q = q_0 q_1$, and $F = f_0 f_1$

This corresponds to two conditioning levels. The first-order WE reduces in fact to two similar equations, one for each level:

Table 4.3 Kinetic model.

	$Y(T, Q)$, $Y(0) = 1$	
$Q\backslash T$	0	1
0	1	1
1	1	0

Table 4.4 Kinetic model (modified).

	$Y(T, Q)$, $Y(0) = 1$	
$Q\backslash T$	0	1
0	1	1
1	1	-1

$$\frac{\partial y_0}{\partial t_0} \oplus q_0 = 0 \tag{4.7}$$

$$y_0(0) = f_0 \tag{4.8}$$

$$\frac{\partial y_1}{\partial t_1} \oplus q_1 = 0 \tag{4.9}$$

$$y_1(0) = f_1 \tag{4.10}$$

The fact that one equation in K is replaced by two equations, one for each level, outlines one of the differences between models in K and S.

The following procedures are suggested by constructions in the categorical frame associated with $SKUP$. This refers to categorical product "$\times$" and coproduct "$\cup$" connecting different conditioning levels. It should be emphasized that they are different from component-wise product $\otimes$ and sum $\oplus$ that refer to elements of the same level. The tensor product connects the levels not the conditions at the same level.

Consider the initial condition:

$$Y(Z,0) = F(Z) = f_0 \times f_1 \tag{4.11}$$

The solution of the model will be

$$Y(T) = y_0 \times y_1 \tag{4.12}$$

The particular case $f_0 = f_1$, $q_0 = q_1$, implies $y_0 = y_1$ (Table 4.5).

This represents the Walsh–Hadamard matrices in DOE. With more coordinates in Y, T, and Z, it is possible to obtain Walsh matrices with 8, 16, 32, and so on elements.

Table 4.6 shows the three levels solution for $Y = y_0 y_1 y_2$ and $T = t_0 t_1 t_2$, a Walsh–Hadamard DOE matrix.

It was considered that $f_0 = f_1 = f_2$, $q_0 = q_1 = q_2$, and this imposes $y_0 = y_1 = y_2$.

Walsh series as solution of differential equations in dyadic field has been obtained by Iordache (1992) and Iordache *et al.* (1993b). Wavelets identified with certain Walsh series and Haar series on the real line have been studied by Lang (1998) in the general frame of formal Laurent series with coefficients "0" or "1."

For the obtained solutions, the tensor product was interpreted as a categorical product, "$\times$". Different classes of design matrices will result if the tensor will be interpreted as a categorical coproduct, "$\cup$".

Table 4.5 Kinetic model.

Product $Y(T) = y_0 \times y_1$			
1	1	1	1
1	−1	1	−1
1	1	−1	−1
1	−1	−1	1

Table 4.6 Kinetic model.

			Product $Y(T) = y_0 \times y_1 \times y_2$				
1	1	1	1	1	1	1	1
1	−1	1	−1	1	−1	1	−1
1	1	−1	−1	1	1	−1	−1
1	−1	−1	1	1	−1	−1	1
1	1	1	1	−1	−1	−1	−1
1	−1	1	−1	−1	1	−1	1
1	1	−1	−1	−1	−1	1	1
1	−1	−1	1	−1	1	1	−1

Table 4.7 Kinetic model.

	Coproduct $Y(T) = y_0 \cup y_1$		
1	1	0	0
1	−1	0	0
0	0	1	1
0	0	1	−1

For two levels, $m = 0$ and $m = 1$, the solution $Y(T)$ may be obtained by stacking matrices as y_0 and y_1 diagonally and completing the missing elements by "0" (see Table 4.7).

This represents an example of coproduct "$\cup$".

The resulting solution shows Haar type functions. A two-setting $\{-1, 1\}$ design is replaced by a three-setting one $\{-1, 0, 1\}$.

The switch from categorical product to coproduct supports the emergence of the new setting "0."

By increasing the number of levels, we may be confronted with an explosive increase in the number of experiments. Switching from categorical product to coproduct allows limiting the size for design matrix.

4.4
"Convection" Model: Latin Squares

Consider now the convective part of the first-order wave equation:

$$\frac{\partial Y}{\partial T} \oplus V \otimes \frac{\partial Y}{\partial Z} = 0 \tag{4.13}$$

The initial condition is

$$Y(Z, 0) = F(Z) \tag{4.14}$$

The operations in K are the sum $\oplus$ and the product $\otimes$ from GF(m).

Table 4.8 Convection model, $m = 3$: $Y(0, Z)$.

Z\V	0	1	2
0	0	0	0
1	1	1	1
2	2	2	2

Table 4.9 Convection model, $m = 3$: $Y(1, Z)$.

Z\V	0	1	2
0	0	1	2
1	1	2	0
2	2	0	1

The general solution of the partial first-order WE is

$$Y(Z, T) = F(Z \oplus (V \otimes T)) \tag{4.15}$$

Consider the initial condition:

$$F(Z) = Z \tag{4.16}$$

This means that at $T = 0$, the output Y of the classification scheme at the distance Z in the scheme is exactly Z. The scheme is one in which each level activates a new difference in properties allowing classification. This kind of initial condition ensures that the wave of the classification process is initiated and is going on.

It results the characteristic

$$Y = Z \oplus (V \otimes T) \tag{4.17}$$

4.4.1
GF(3) Solution

For $T = 0$ the solution Y is shown in Table 4.8; for $T = 1$ the solution is shown in Table 4.9; and for $T = 2$ the solution is shown in Table 4.10.

Table 4.10 Convection model, $m = 3$: $Y(2, Z)$.

Z\V	0	1	2
0	0	2	1
1	1	0	2
2	2	1	0

Table 4.11 Concatenated solutions, $m = 3$.

000	012	021
111	120	102
222	201	210

Table 4.12 Pasting down columns, $m = 3$.

F0	F1	F2
0	0	0
1	1	1
2	2	2
0	1	2
1	2	0
2	0	1
0	2	1
1	0	2
2	1	0

There is a relation between different solutions of the first-order wave equation and DOE matrices.

For different values of T, $T = 1$ and 2, one obtains different (3×3) Latin squares L. Latin squares' close association with DOE is well known (Hedayat, Sloane and Stufken, 1999).

The following procedures to obtain DOE are suggested by universal constructions in categorical frame. There are several DOE to be obtained by combining the solutions obtained for different values of T.

Superposing by concatenation the elements of the Tables 4.8–4.10 the Table 4.11 will result.

Pasting down the three-digit numbers from Table 4.11, column after column, the Table 4.12 is obtained. It is a well-known DOE with nine experiments for three factors.

Columns in Table 4.12 are orthogonal. Each column corresponds to first-order wave equation solutions at different velocities, V. Associating one supplementary digit for each column in Table 4.11 will result in the four-digit numbers as in Table 4.13. Here, the index (0) corresponds to the first column in Table 4.10, (1) to the second column, and (2) to the third column.

Table 4.13 Indexed concatenated solutions, $m = 3$.

(0)000	(1)012	(2)021
(0)111	(1)120	(2)102
(0)222	(1)201	(2)210

Table 4.14 Convection model, $m = 4$: $Y(0, Z)$.

Z\V	0	1	2	3
0	0	0	0	0
1	1	1	1	1
2	2	2	2	2
3	3	3	3	3

Table 4.15 Convection model, $m = 4$: $Y(1, Z)$.

Z\V	0	1	2	3
0	0	1	2	3
1	1	0	3	2
2	2	3	0	1
3	3	2	1	0

The resulting four-digit numbers from Table 4.13 correspond to columns of well-known orthogonal design with nine experiments and four factors (Taguchi, 1986; Hedayat, Sloane and Stufken, 1999).

It is the same array as the OA shown in Table 4.12 except that it has a supplementary column with four identical entries, 0, 1, or 2.

Concatenation and pasting down operations are, in the main, close to coproduct "∪" type of operation in categorical frame.

4.4.2
GF(4) Solution

The operations in the model are the component-wise sum ⊕ and product ⊗ from GF(4).

The solutions for different T are presented in what follows.

Tables 4.14–4.17 show the case $T = 0$, $T = 1$, $T = 2$, and $T = 3$.

Table 4.16 Convection model, $m = 4$: $Y(2, Z)$.

Z\V	0	1	2	3
0	0	2	3	1
1	1	3	2	0
2	2	0	1	3
3	3	1	0	2

Table 4.17 Convection model, $m = 4$: $Y(3, Z)$.

Z\V	0	1	2	3
0	0	3	1	2
1	1	2	0	3
2	2	1	3	0
3	3	0	2	1

For different values of T, $T = 1, 2$, and 3, one obtains different (4×4) Latin squares, L.A superposition by concatenation of solutions obtained for $T = 0$, $T = 1$, $T = 2$, and $T = 3$ may be considered.

Table 4.18 results in this case.

Pasting down the four-digit numbers from Table 4.18, column after column results in Table 4.19.

Columns may be associated with factors, for instance, F0, F1, F2, and F3. Columns are orthogonal.

Table 4.18 Concatenated solutions, $m = 4$.

0000	0123	0231	0312
1111	1032	1320	1203
2222	2301	2013	2130
3333	3210	3102	3021

Table 4.19 Pasting down columns, $m = 4$.

F0	F1	F2	F3
0	0	0	0
1	1	1	1
2	2	2	2
3	3	3	3
0	1	2	3
1	0	3	2
2	3	0	1
3	2	1	0
0	2	3	1
1	3	2	0
2	0	1	3
3	1	0	2
0	3	1	2
1	2	0	3
2	1	3	0
3	0	2	1

Table 4.20 Indexed concatenated solutions, $m = 4$.

(0)0000	(1)0123	(2)0231	(3)0312
(0)1111	(1)1032	(2)1320	(3)1203
(0)2222	(1)2301	(2)2013	(3)2130
(0)3333	(1)3210	(2)3102	(3)3021

There are different other DOE that may be obtained by combining the WE solutions for different T.

Associating one supplementary digit for each column in Table 4.18, five-digit numbers will result, for instance, (0)0000, (0)1111, (1)0123, (3)1203, and so on (see Table 4.20).

The resulting five-digit numbers correspond to columns of well-known orthogonal design (16 experiments, 5 factors) (Taguchi, 1986; Hedayat, Sloane and Stufken, 1999).

It is the same array as the OA shown in Table 4.19 except that has a supplementary column with four identical entries, 0, 1, 2, or 3 associated with columns.

The previously obtained matrices are linked to the tensor product interpretation as coproduct "∪".

Obviously, using the tensor product as categorical product will offer other class of solutions, asking for significantly more experiments. It is known that the product of two Latin squares gives another Latin square. For instance, the Kronecker product of two (3×3) Latin squares gives a (9×9) Latin square. Moreover, the Kronecker product of a difference scheme as that resulting from the pasting down, described here, to an orthogonal array gives another orthogonal array (Hedayat, Sloane and Stufken, 1999). This means that switching from categorical product to categorical coproduct way maintains orthogonality.

In fact, the switch in both directions corresponds to categorical natural transformations.

As far as the kinetic types of solutions of the WE are concerned, increasing the number of levels will increase dramatically the number of experiments if the categorical product way is followed. The transition from categorical product to coproduct allows controlling the size of the design of experiment matrix. The reverse movement, from coproduct to categorical product, is also beneficial since it enriches the search space and avoids missing significant factors in problem solving.

4.5
Spectral Analysis: Correlation

The reason for performing spectral analysis of a pattern is to determine the power transported by each harmonic. The Walsh–Fourier analysis based on Walsh–Hadamard functions is widely applied in information processing (Elliot and Rao, 1982).

Similar analysis can be performed for Haar functions. Notice that at the same number of levels, it is computationally advantageous to use Haar functions instead of Walsh functions (Khuri, 1994).

The DOE factors, corresponding to orthogonal columns in DOE matrices, are the harmonics here. Denote by $W(T, V)$ such columns. The result $Y(T)$ of a DOE experiment may be, for instance, the percentage of failures or the records of the presence of failure.

The real Fourier spectrum (or other correlation index) allows fast recognition of patterns.

The vector $Y(T)$ can be expressed as Fourier series of solutions $W(T, V)$ functions.

$$Y(T) = \sum_V q_V W(T, V) \tag{4.18}$$

The coefficient q_V depends on the contribution of V to the object $Y(T)$. The spectrum q_V is given by

$$q_V = \sum_T Y(T) W(T, V) \tag{4.19}$$

Recall that both the product and the sum have been considered here in real field.

It is possible to use not only real but also $GF(m)$ operations for the sum $\oplus$ and the product $\otimes$ in correlation.

In that case:

$$Y(T, V) = \sum_{\oplus v} q_v \otimes W(T, V), \tag{4.20}$$

$$q_V = \sum_{\oplus T} Y(T) \otimes W(T, V) \tag{4.21}$$

Correlation plays a role comparable to probabilities in real-field frames.

The more correlated factor or the more correlated experiment will be selected as the more important for the next step of search. Suppose, for example, the result of a DOE experiment is the percentage of failure or the records of the presence of failure. Correlation means that to the same digit in factor description, for example, to"1" corresponds the same trend for percentage of failure, for example, "high". If for the same digit, "1", the percentage of failure is sometimes high and sometimes low, that is, if the factor and the results do not show a similar trend, it shows that the factor is not the important one for the studied failure.

The first-order WE allows analyzing the data in a way that differs essentially from conventional Fourier series expansions for real field. For Fourier series the accuracy increases as the number of terms in expansion increases. In the proposed method, the classification rate Q or the velocity V may be imposed, and the number of scales or levels m allowing to improve the fitting may be established by comparison of models resulting for different m with the experiment. Starting from a given number of levels m, the expansion diverges and this may be interpreted as a limit in the number of structured conditioning levels. The aim in this case is not the optimal fitting but the labeling of the compound process according to its organization in levels and to limit the number of experiments at an acceptable fitting.

5
Informational Analysis: EDOE Matrices

5.1
Walsh–Hadamard Matrices and Latin Square Designs

The evolvable designs of experiments (EDOEs) construction may start with matrices resulting as solutions of the partial first-order wave equation (WE) over finite fields. Some of these matrices are similar to conventional fractional factorial designs. Restricting the start of EDOE analysis to WE generated and information significant design of experiment (DOE) matrices introduces consistence into problem solving by driving the analyst to begin from similar building blocks. By confronting and embedding reality input, a large variety of EDOE networks will result.

The Walsh–Hadamard (WH) designs are screening experiments. They can determine the individual effects for each variable on a response, but these types of DOE cannot detect interactions. To account for factor interactions, the problem solving process needs to modify the W and L matrices adaptively. Significant in this case is the strategy of advancement of the design steps across the problem framework. This makes specific use of the technical expertise related to the studied domain. The objective may be to select for each DOE step the matrices offering maximum information to drive the whole framework toward specified thresholds of production of information and so forth. In such cases, the informational entropy associated with the DOE matrix and EDOE frame should be evaluated.

Observe that the so-called robust designs or Taguchi designs, such as L4, L8, L16 (15 factors and 2 levels), and so on, are particular Walsh functions, Walsh–Hadamard, or Walsh–Paley (WP), designs that are slightly modified. They have been discovered long time before robust design methodology. The robust design uses the Walsh matrices without the first column. The designs L4, L8, L16, and so on denote the robust design orthogonal Walsh-type matrices. They correspond to 4, 8, and 16 runs and to 3, 7, and 15 factors, respectively. In fact, L4 results from WP4, L8 from WP8, L16 from WP16, without the first column.

Evolvable Designs of Experiments: Applications for Circuits. Octavian Iordache
Copyright © 2009 WILEY-VCH Verlag GmbH & Co. KGaA, Weinheim
ISBN: 978-3-527-32424-8

5.2
Classification Procedures: Informational Criteria

Each line in the matrix of DOE corresponds to an experiment, that is, to specific settings and each column to a factor. Statistical DOE makes use of such matrices and the corresponding responds to establish the effects of different factors (focus on columns). In EDOE method, a supplementary objective is the classification of experiments or samples (focus on rows).

The classification method parallels and simplifies partition in classes for stochastic processes, replacing transition probabilities by similarities.

The procedure is as follows: let $i = [i_1, \ldots, i_k, \ldots]$ and $j = [j_1, \ldots, j_k, \ldots]$ be two row vectors, that is, two runs. To any pair of vectors, the weighted similarity r_{ij} is associated.

$$r_{ij} = \sum_k t_k (a_k)^k \tag{5.1}$$

With $0 < a_k < 1$, a constant $t_k = 1$ if $i_k = j_k$, and $t_k = 0$ if $i_k \neq j_k$. One use $a_k = 0.5$, corresponding to GF(2) here. This means that the first factor has the weight 0.5, the next 0.25, and the next 0.125, and so on. This is the GF(2) frame situation for any coordinate in vectors.

Taking into account the easiness of calculus in GF(2), similarities may replace and remarkably simplify the real-field defined probability calculus.

To any matrix of design corresponds a similarity matrix $R = [r_{ij}]$. This is the analogous of transition probabilities matrix associated with Markov stochastic processes.

The stabilization procedure allowing partition in disjoint classes is mandatory. To obtain the stable matrix, denoted by $R(n) = [r_{ij}(n)]$, the composition rule $R(n) = R(n-1) o R$ is applied starting from $R(2) = R o R$ toward the stable matrix $R(n) = R(n+1) = R(n+2) = \cdots$. For two similarity matrices $R = [r_{ij}]$ and $Q = [q_{ij}]$, the composition rule "o" is defined by

$$(R o Q)_{ij} = \max_k [\min(r_{ik}, q_{kj})] \tag{5.2}$$

This gives the ijth element of the composed matrix. The classification algorithm is as follows: two experiments i and j are assigned to the same class at the classification level $T, 0 \leq T \leq 1$, if their similarity in the stable matrix is larger than T, that is, if $r_{ij}(n) > T$.

In a more general way, it is possible to start by introducing a real-valued distance d_{ij} between any two vectors i and j.

The Minkowski distance is

$$d_{ij} = \left(\sum_k (i_k - j_k)^k \right)^{1/k} \tag{5.3}$$

If $k = 2$, this gives the Euclidean distances. The Manhattan or city-block distance is also of interest. This is defined as

$$d_{ij} = \sum_k |i_k - j_k| \tag{5.4}$$

It is equivalent to the number of settings that are different in the experiments i and j.

The Hamming distance is defined by

$$d_{ij} = \sum_k \text{XOR}(i_k, j_k) \tag{5.5}$$

XOR is the exclusive or logical function. For Boolean vectors, the Manhattan and the Hamming distances are equivalent.

The Tanimoto distance is defined by

$$d_{ij} = \sum_k \frac{\text{AND}(i_k, j_k)}{\text{OR}(i_k, j_k)} \tag{5.6}$$

AND and OR are logical functions.

A specific distance is

$$d_{ij} = \sum_k t_k (a_k)^k \tag{5.7}$$

With $0 < a_k < 1$, a constant, $t_k = 1$ if $i_k = j_k$, $t_k = 0$ starting from the first k with $i_k \neq j_k$.

If the comparison between i and j stops at the first difference in coordinates are detected and $a_k = 0.5$, the distance is ultrametric.

Distances are measures of dissimilarity. Once the distance is established, similarity indices may be obtained using

$$r_{ij} = \frac{1}{(1 + d_{ij})} \tag{5.8}$$

Observe that $r_{ij} < 1$ and $r_{ii} = 1$.

Other definitions for similarities are

$$r_{ij} = 1 - d_{ij} \tag{5.9}$$

$$r_{ij} = 1 - \frac{d_{ij}}{\max(d_{ij})} \tag{5.10}$$

The elements of the categorical frame $SKUP$ may be easily revealed in such classification procedures. The category S corresponds to the similarity matrices for experiments, while the category K corresponds to the similarity matrices for the classes or lumps.

S and K are categories since the identity, associativity, and commutativity of the morphisms hold in S and in K. The problem is to define functors U that will be a natural way to transform the category K of lumped matrices in the category S of similarity matrices and functors P that will allow to transform S back to K. The study of classification schemes outlined the advantage of fuzzy measures replacing probabilities in defining the functors U and P (Iordache, Bucurescu and Pascu, 1990; Iordache, Corriou and Tondeur, 1993c).

5.3
Informational Entropy and Distances

An open problem is to select the set of significant experiments associated with a DOE matrix. The solution takes into account informational criteria. With any matrix of design, a similarity matrix R is associated, and with this informational entropy:

$$H(R) = -\sum r_{ij} \ln r_{ij} - \sum (1-r_{ij}) \ln (1-r_{ij}) \tag{5.11}$$

This expresses the quantity of information associated with the matrix of design.

The defined entropy is a measure of the imprecision in classifying the experiments. This entropy approaches the maximum value for equal similarities, that is, 0.5 (Iordache, Bucurescu and Pascu, 1990). The higher entropy is an objective since it offers significance to experiments to be performed.

To have a choice between more possible matrices of DOE, M, a matrix can be utilized, say Q, as reference and the distance between M and Q as an optimality criterion.

Suppose, to the design matrices M and Q correspond, respectively, the similarity matrices $R = [r_{ij}]$ and $S = [s_{ij}]$. The distance is

$$DD(M, Q) = -\sum r_{ij} \ln \left(\frac{r_{ij}}{s_{ij}}\right) - \sum (1-r_{ij}) \ln \left(\frac{1-r_{ij}}{1-s_{ij}}\right) \tag{5.12}$$

This informational distance helps in selecting the design matrices.

5.4
Adaptability in Classification

The DOE matrices need to be trained to ensure the right hierarchical ranking of factors. Training is based on "known as good" diagnosis, known good solutions, and quality classifications as established by practical expertise or on the basis of different tests.

The learning consists in seeking for a matrix that imposes a classification that is as similar as possible to the "good" classification. Learning is viewed as a matter of classifying and grouping the elementary units into a smaller number of richer, more densely packed information.

The learning procedures are based on operations as follows:

- Changing the number of factors

If a factor is not significant, it may be discarded. Keeping one or more columns of an orthogonal array empty, DOE does not lose orthogonality of a matrix experiment.

- Changing the hierarchical position of factors
- Changing the number of digits affected to a factor.

This is achieved by splitting the column for the factor.
An opposite procedure is that of merging the columns.

- Changing the weights a_k of factors.

The weights can be adjusted in a trial-and-error fashion until a maximally effective combination of the weighted features is achieved.

- Changing the number of setting levels of factors.

Merging column with or without keeping the orthogonality could do this operation.

The utilized algorithms are based on informational criteria as the distance DD between two matrices. For instance, if the classification of some samples in the "0" class of quality does not correspond to expert criteria, it is possible to modify the order of factors to ensure the supposed good classification. In this way, the nonlinear interaction between factors is taken into account.

5.5
Informational Results

Wn, m, s denotes the Walsh DOE matrix with n rows (runs or experiments), m columns (factors), and s maximum settings (levels for parameter values). Frequently, n varies from 3 to 16, m varies from 2 to 15, while the number of settings s is 2. DOE with more than two settings of each factor have been considered too.

The similarity with $a_k = 0.5$ and the entropy as defined by Equation 5.11 is calculated in what follows. These choices limit the generality of the numerical simulation results.

5.5.1
Proposition 1

Consider particular Walsh systems, the Walsh–Paley matrices, Walsh–Hadamard matrices, as shown in Table 5.1. Denote by $n = 2^k$, the number of experiments, that is, runs and the number of factors by m.

The ratio $H(WP2^n)/H(WP2^{n-1})$ tends to 4 as n increases.

The entropy associated with WP4 is $H(WP4) = 7.542$. Also, $H(WP8) = 30.371$ and $H(WP16) = 121.710$.

As expected, more experiments are able to increase the significance of the DOE.

Notice that these types of Walsh matrices are resulting by successive application, level after level, of the categorical product for WE solution.

Table 5.1 Matrices WP8 and WH8.

			WP8								WH8				
1	1	1	1	1	1	1	1	1	1	1	1	1	1	1	1
1	1	1	1	−1	−1	−1	−1	1	−1	1	−1	1	−1	1	−1
1	1	−1	−1	1	1	−1	−1	1	1	−1	−1	1	1	−1	−1
1	1	−1	−1	−1	−1	1	1	1	−1	−1	1	1	−1	−1	1
1	−1	1	−1	1	−1	1	−1	1	1	1	1	−1	−1	−1	−1
1	−1	1	−1	−1	1	−1	1	1	−1	1	−1	−1	1	−1	1
1	−1	−1	1	1	−1	−1	1	1	1	−1	−1	−1	−1	1	1
1	−1	−1	1	−1	1	1	−1	1	−1	−1	1	−1	1	1	−1

5.5.2
Proposition 2

Consider the number of experiments, $n = 2^k$, fixed. As m increases from $m = 1$ to $m = 2^k$, the entropy of WP matrices decreases and evolves toward minimal informational entropy.

$$\frac{\Delta H}{\Delta m} \leq 0 \tag{5.13}$$

Some examples are presented in Tables 5.2–5.4.

Observe that the production of entropy tends to zero as m tends to $n - 1$.

Since the defined entropy is a measure of the imprecision in classifying the experiments, it results that for fixed number of experiments n, it would be beneficial to have fewer factors m to be tested. The number of factors or experiments included in a DOE may be optionally selected to ensure an imposed minimal variation of informational entropy for new factors.

5.5.3
Proposition 3

Consider the number of factors $m = 2^k$ constant. As n increases, the informational entropy increases too (Tables 5.5 and 5.6).

$$\frac{\Delta H}{\Delta n} \geq 0 \tag{5.14}$$

Table 5.2 WP4 entropy variation for n constant.

m	1	2	3	4
$H(\text{WP4})$	8.318	7.795	7.542	7.542

Table 5.3 WP8 entropy variation for n constant.

m	1	2	3	4	5	6	7	8
$H(\text{WP8})$	38.816	35.677	33.182	32.038	30.867	30.537	30.371	30.371

Table 5.4 WP16 entropy variation for n constant.

m	1	2	3	4	5	6	7	8
$H(\text{WP16})$	166.365	151.704	138.755	131.892	125.694	123.437	122.216	121.894
m	9	10	11	12	13	14	15	16
$H(\text{WP16})$	121.381	121.209	121.122	121.101	121.079	121.073	121.710	121.710

Table 5.5 WP4 entropy variation for m constant.

n	1	2	3	4
H (WP4)	0	1.125	3.771	7.542

Table 5.6 WP8 entropy variation for m constant.

n	1	2	3	4	5	6	7	8
H (WP8)	0	0.468	2.486	4.973	10.079	15.653	22.778	30.371

This means that for an imposed number of factors to be tested, the WP design offers more imprecision in classifying and more significance to the test if the number of trials increases.

In experimental activity, there is a natural trend to obtain maximum significance, that is, maximum entropy with limited imposed resources. The strategy depends on similarity definition and on the set of matrices used in building EDOE.

One experimental strategy is to fill up DOE matrices with new experiments, that is, new rows, since the goal is to increase entropy. If the system is not in the state of maximum entropy, it would be necessary to perform new experiments at the same level to produce more information. If the production of informational entropy by specific new experiments becomes lower than an imposed threshold, new levels, that is, new factors should be involved, and this would help increase of the total system entropy.

Informational results, as shown in Table 5.7, may be of help in initial choice of matrices.

Notice that the similarity with $a_k = 0.5$ and the entropy, as defined by Equation 5.11, is utilized for Table 5.7.

The informational entropy behavior is related also to the fact that as m increases the weight of new factors decreases. Other definitions of similarity will give different numerical results.

Table 5.7 Entropy for matrix $WHn,m,2$.

m	2	3	4	5	6	7	8
n	4	4	8	8	8	8	12
(n, m)	(4,2)	(4,3)	(8,4)	(8,5)	(8,6)	(8,7)	(12,8)
H	7.27	7.27	36.25	36.70	36.56	36.49	81.71
m	9	10	11	12	13	14	15
n	12	12	12	16	16	16	16
(n, m)	(12,9)	(12,10)	(12,11)	(16,12)	(16,13)	(16,14)	(16,15)
H	81.66	81.67	81.66	136.87	136.86	136.87	136.87

5.6
Relation with Thermodynamics

The correspondence between Boltzmann's statistical interpretation of physical entropy as disorder and Shannon's formulation of variety and imprecision as informational entropy suggests comparing the informational results with those from thermodynamics.

The Proposition 3 and the associated Tables 5.5 and 5.6 refer to the increase of entropy with the number of experiments that is, in time and is analogous to the second law of thermodynamics.

However, some EDOE calculations proved that it is possible that the information entropy decreases with new experiments, apparently contradicting the second law of thermodynamics. First, it should be emphasized that the above-calculated entropies are for the conditions space K; that is, they are informational entropies. The EDOE includes, in some sense, the condition space K and also the real state system S, with its entropy.

The apparently inconsistent behavior of informational entropy in K has been signaled by several authors as, for instance, Iordache, Corriou and Tondeur (1993c) in the study of separation sequences, Parunak and Brueckner (2001) in the study of multiagent systems, Adami (2002) in the study of the "physical complexity," and Mnif and Muller-Schloer (2006) in the study of emergence for "organic computation."

This discrepancy may be explained in all these cases in terms of Kugler and Turvey (1987) model. Self-organization and loss of entropy occur at the so-called macrolevel, that is, in K, while the system dynamics at the real microlevel, that is, in S generates increasing disorder and serves as entropy "sink." This permits the total entropy of the overall system, that is, in K and S to increase, while allowing self-organization to emerge in K (Parunak and Brueckner, 2001).

The information distance, that is, the conditional entropy falls observed in the sequencing study (Iordache, Corriou and Tondeur, 1993c) was interpreted as the emergence of a new classification or separation scheme at the macrolevel. In that study, the microlevel, that is, the real space S was represented by the random process generating the real-valued weights, differing from the exact powers of 0.5. The entropy associated with the random weights in S was able to support the emergence of new separation schemes after a few numerical steps, in K.

Adami (2002) simulations also show domains of complexity declines in time. Physical complexity, as introduced by Adami, assigns low complexity to both ordered and random systems and high complexity to those between them. Physical complexity measures the amount of information that a system stores in its space of conditions K about the environment S in which it evolves. Thus, evolution increases the amount of knowledge a population accumulates about its niche. Since entropy is a measure of potential information, biological evolution leads to decrease of entropy. Dynamic environment, that is, S continually creates the problem to be solved. To survive in the environment means to solve problems, the solutions being embodied knowledge. By thermodynamic reasoning, it is possible to identify differentiation from environment, dissipative structures, or other far-from-equilibrium structures. The basic idea for physical complexity definition was to relate complexity to system's information about its environment. In this way, the informational and thermodynamic approaches to complexity may be closely related.

Moreover, the study of both types of entropy allows detecting emergence.

Tables 5.5 and 5.6 show that the increases in entropy from one experiment to another start with a minimum increase of entropy, for the first step from $n = 1$ to $n = 2$. Then, the increase of entropy in each step increases with n. This behavior may be explained in terms of the approach to nonequilibrium in statistical thermodynamics. The first experiment activates only one of the possible levels, that is, only one scale contributing to the entropy production. This situation is parallel to the minimum production of entropy (Prigogine, 1980). As more experiments are performed and n increases, other scales of entropy production may be activated and the system tends toward the maximum production of entropy (Dewar, 2003).

As a significant difference from physical systems, in EDOE the application of informational theorems appears to be context dependent and limited.

For complexity conditions, the evolvability rather than optimization is the suitable measure of the methodology potential to solve problems. In numerous practical situations, it may be impossible to achieve the optimum due to the restrictions in time and resources. Complex systems are grounded on evolvability rather than on optimization criteria.

5.7
Ranking, Discarding, and Replication of the Columns

Several procedures to modify DOE matrices will be presented in what follows.

The matrix designs that are orthogonal and balanced allow seeing the true effects of the inputs independent of each other. However, the orthogonality is not always required nor recommended. The first-order wave equations over finite fields introduced in the Chapter 4 generate standard L and W matrices. From these, several variants may result, as shown next. It is possible to make use of standard matrices but change the order and the number of columns.

The hierarchy of factors imposes the change. Experiments aiming to analyze the effects of intermediary number of factors as $n - 3, n - 2, n - 1$ factors, using n trials, may be based on incomplete forms of the n trial matrices.

Denote by 1–8 the successive columns in WP8, a Walsh–Paley type of matrix. The entropy associated with different set of columns in the specified order is included in Table 5.8.

Observe that for any number of factors, that is, of columns, there is a specific succession offering more informational entropy. This challenges some of the

Table 5.8 WP8 entropy variation for $n = 8$ and different sequences of columns.

Selected columns	4,5	4,5,6	4,5,6,3	4,5,6,3,2	4,5,6,3,2,7	4,5,6,3,2,7,8
Entropy, H	33.584	36.254	36.254	36.699	36.562	36.492

selections arbitrarily operated in robust design for *k* factors and eight trials (Lochner and Matar, 1990).

It should be noted that contrary to EDOE approach, for robust design method as for conventional DOE, the order of columns is not an issue. The factors may be assigned more or less arbitrarily to the columns. Table 5.8 shows that for eight runs and two settings, the sequence 4,5,6,3,2,7,8 allows running, for each experiment stage, close to maximum entropy. This is important since this represents the order of questioning the physical system and the ranking of decisions in problem solving.

Recall that the entropy has been calculated assuming that it is a hierarchical order of significance, for factors relative to the problem to be solved. For some type of distances introduced in this part, the sequence of the columns proves to be of low significance. Matrices such as Wn,m,s are expected to be utilized for $n-1$ factors. Such matrices can also be operated to study less than $n-1$ factors, if necessary. Some of the columns are not assigned to factors. This discarding is not at random but may be done according to informational criteria too. Starting from the completed matrices, columns are discarded one by one to meet the request of minimum loss of informational entropy.

Sometimes, it is necessary to ensure enhanced significance to a factor. This is achieved by doubling the column associated with that factor. The DOE loses its orthogonality but during classification by weighted distances, the double digits outline the contribution of the factors.

5.8
Lumping and Splitting Columns

Lumping is a modification of a set of array columns to allow more settings of a factor to be investigated. A weight is associated with any column. For instance, if three columns are selected to be lumped together, the setting of the first column is multiplied by $4 = 2^2$, the second by $2 = 2^1$, and of the third by $1 = 2^0$. The new column is obtained by summing on rows. Consider, for example, the matrix W4,3,2a in Table 5.9. One choice is to lump columns 1 and 2, giving the weight $2 = 2^1$ to the first column digits and $2 = 2^0$ to the second column digits. The last column is not modified. It results in the matrix W4,2,4.

Four settings may be assigned to the first factor, now corresponding to $+3$, $+1$, -1 and -3. The new column is orthogonal relative to the resting one columns but not to

Table 5.9 Matrices W4,3,2a and W4,2,4.

W4,3,2a			W4,2,4	
+1	+1	+1	+3	+1
+1	−1	−1	+1	−1
−1	+1	−1	−1	−1
−1	−1	+1	−3	+1

that included in its construction. To maintain the orthogonality, it is necessary to select any column just one time. The usually neglected first column from Walsh type of matrices with identical settings now became useful. The above procedure is automatic and offers a high number of matrices as per the number of settings of interest.

The splitting of columns was also considered.

As before, the elements of the columns are interpreted as dyadic numbers. For instance, the setting 1 is replaced by the vector (0,1), the setting 2 by (1,0), the setting 3 by (1,1), and so on, as in a dyadic expansion. Then the column is divided into two columns. In this way, two-setting factors may be considered together with three-setting factors. Splitting allows fewer settings of a factor to be investigated. A similar effect is obtained by grouping settings and may be performed by replacing settings 1 or 2 by 2 only and in the same way, settings -1 or -2 by -2 only.

5.9
Juxtaposing and Cutting

Modified W and L type of matrices have been obtained by pasting right or pasting down different model generated elementary matrices. To paste right, it is necessary to have the same number of experiments n. To paste down, the same number of factors m is required.

Concatenation procedures in generating DOE matrices have been presented before.

Cutting in some case is unavoidable since some runs may be missed in industrial experiments. There exist cuttings that maintain orthogonality for columns that are different. Taking into account that Walsh–Hadamard matrices can be obtained by replication of lower order matrices, it is clear that such matrices may be segregated in orthogonal matrices (Elliot and Rao, 1982). The procedure reverses the construction process shown in Tables 4.5 and 4.6. The coproduct procedure as shown in Table 4.7 may limit the number of experiments.

The methods presented above to generate matrices have something in common with neural networks (NNs) and genetic algorithm strategies. It should be emphasized that standard neural networks and genetic algorithms-based neural learning methods are computationally expensive. Thousands of iterations are common to back propagation methods and thousands of generations with hundreds of individuals in the populations are common to genetic algorithms.

It was observed that neural networks and genetic algorithms are impractical for some complex industry failures when it is expected to quickly learn and solve the problem in real time frame. An example of neural network frame in which a weight was associated to any column was presented by Iordache *et al.* (1993a). Significant difference between EDOE and neural learning tools such as neural network or genetic algorithm is that the EDOE is based on model-generated matrices and not on digital vectors generation selection and evolution by random search. The pasting of the component of a vector and the splitting of a column are based on entropy calculus in EDOE not on random grouping and breaking processes having no relation with the studied problem and low chance to find optimality. It was recognized that genetic algorithms work well in some cases and not in others, but it is not yet clear why this happens. Most likely, EDOE methodology may offer a partial answer to the deception associated with genetic algorithm applications (Whitley, 1991).

5.10
Tables of DOE Matrices

Few examples of DOE matrices obtained by the described procedures are presented next (Tables 5.10–5.16).

Table 5.10 Matrices W4,2,2a; W4,2,4a; W4,3,2a.

W4,2,2a		W4,2,4a		W4,3,2a		
1	1	3	1	1	1	1
1	−1	1	−1	1	−1	−1
−1	+1	−1	−1	−1	1	−1
−1	−1	−3	1	−1	−1	1

Table 5.11 Matrices W8,4,2a; W8,3,4a; W8,4,4a; W8,4,4b.

W8,4,2a				W8,3,4a			W8,4,4a				W8,4,4b			
1	1	1	1	3	1	1	3	3	3	3	3	3	3	3
1	−1	−1	1	1	−1	1	1	−1	1	−1	1	−1	1	1
−1	1	1	−1	−1	1	−1	−1	1	1	−1	−1	1	1	1
−1	−1	−1	−1	−3	−1	−1	−3	−3	3	3	−3	−3	3	3
−1	1	−1	1	−1	−1	1	−1	−1	−1	−1	−1	−1	−1	1
−1	−1	1	1	−3	1	1	−3	3	−3	3	−3	3	−3	3
1	1	−1	−1	3	−1	−1	3	−3	−3	3	3	−3	−3	3
1	−1	1	−1	1	1	−1	1	1	−1	−1	1	1	−1	1

Table 5.12 Matrices W8,4,4c; W8,4,4d; W8,4,8a; W8,5,2a.

W8,4,4c				W8,4,4d				W8,4,8a				W8,5,2a				
3	3	3	1	3	1	1	1	−4	1	1	1	1	1	1	1	1
1	−1	1	−1	1	−1	1	1	−3	1	−1	−1	1	−1	−1	1	1
−1	1	1	−1	−1	1	−1	1	−2	1	−1	−1	−1	1	1	−1	1
−3	−3	3	1	−3	−1	−1	1	−1	1	1	1	−1	1	−1	1	−1
−1	−1	−1	−1	−1	−1	1	−1	1	−1	1	−1	−1	1	−1	1	−1
−3	3	−3	1	−3	1	1	−1	2	−1	−1	1	−1	−1	1	1	−1
3	−3	−3	1	3	−1	−1	−1	3	−1	−1	1	1	1	−1	−1	−1
1	1	−1	−1	1	1	−1	−1	4	−1	1	−1	1	−1	1	−1	−1

Table 5.13 Matrix W12,8,2a.

W12, 8,2a							
1	1	1	1	1	1	1	1
1	1	1	1	1	-1	-1	-1
1	1	-1	-1	-1	1	1	1
1	-1	1	-1	-1	1	-1	-1
1	-1	-1	1	-1	1	-1	-1
1	-1	-1	-1	1	-1	-1	1
-1	1	-1	-1	1	1	-1	-1
-1	1	-1	1	-1	-1	-1	1
-1	1	1	-1	-1	-1	1	-1
-1	-1	-1	1	1	1	1	-1
-1	-1	1	-1	1	-1	1	1
-1	-1	1	1	-1	1	-1	1

Table 5.14 Matrix W16,12,2a.

W16,12,2a											
1	1	1	1	1	1	1	1	1	1	1	1
1	1	1	1	1	1	1	-1	-1	-1	-1	-1
1	1	1	-1	-1	-1	-1	1	1	1	1	-1
1	1	1	-1	-1	-1	-1	-1	-1	-1	-1	1
1	-1	-1	1	1	-1	-1	1	1	-1	-1	1
1	-1	-1	1	1	-1	-1	-1	-1	1	1	-1
1	-1	-1	-1	-1	1	1	1	1	-1	-1	-1
1	-1	-1	-1	-1	1	1	-1	-1	1	1	1
-1	1	-1	1	-1	1	-1	1	-1	1	-1	1
-1	1	-1	1	-1	1	-1	-1	1	-1	1	-1
-1	1	-1	-1	1	-1	1	1	-1	1	-1	-1
-1	-1	1	1	-1	-1	1	1	-1	-1	1	1
-1	-1	1	1	-1	-1	1	-1	1	1	-1	-1
-1	-1	1	-1	1	1	-1	1	-1	-1	1	-1
-1	-1	1	-1	1	1	-1	-1	1	1	-1	1
-1	-1	1	-1	1	1	-1	-1	1	1	-1	1

Table 5.15 Matrices L9,4,3a; L9,4,3b;L9,3,9a; L9,3,9b.

L9,4,3a				L9,4,3b				L9,3,9a			L9,3,9b		
-1	-1	-1	-1	-1	-1	-1	-1	-4	-1	-1	-4	-1	-1
-1	0	0	0	-1	0	0	1	-3	0	0	-3	0	1
-1	1	1	1	-1	1	1	0	-2	1	1	-2	1	0
0	-1	0	1	0	-1	0	0	-1	0	1	-1	0	0
0	0	1	-1	0	0	1	-1	0	1	-1	0	1	-1
0	1	-1	0	0	1	-1	1	1	-1	0	1	-1	1
1	-1	1	0	1	-1	1	1	2	1	0	2	1	1
1	0	-1	1	1	0	-1	0	3	-1	1	3	-1	0
1	1	0	-1	1	1	0	-1	4	0	-1	4	0	-1

Table 5.16 Matrices L16,5,4a; L16,5,4b; L16,4,13a.

L16,5,4a					L16,5,4b					L16,4,13a			
−2	−2	−2	−2	−2	−2	−2	−2	−2	−2	−6	−2	−2	−2
−2	−1	−1	−1	−1	−2	−1	−1	1	2	−5	−1	−1	−1
−2	1	1	1	1	−2	1	1	2	−1	−3	1	1	1
−2	2	2	2	2	−2	2	2	−1	1	−2	2	2	2
−1	−2	−1	1	2	−1	−2	−1	−1	−1	−4	−1	1	2
−1	−1	−2	2	1	−1	−1	−2	2	1	−3	−2	2	1
−1	1	2	−2	−1	−1	1	2	1	−2	−1	2	−2	−1
−1	2	1	−1	−2	−1	2	1	−2	2	0	1	−1	−2
1	−2	1	2	−1	1	−2	1	1	1	0	1	2	−1
1	−1	2	1	−2	1	−1	2	−2	−1	1	2	1	2
1	1	−2	−1	2	1	1	−2	−1	2	3	−2	−1	2
1	2	−1	−2	1	1	2	−1	−2	1	4	−1	−2	1
2	−2	2	−1	1	2	−2	2	2	2	2	2	−1	1
2	−1	1	−2	2	2	−1	1	−1	−2	3	1	−2	2
2	1	−1	2	−2	2	1	−1	−2	1	5	−1	2	−2
2	2	−2	1	−1	2	2	−2	1	−1	6	−2	1	−1

6
EDOE Methodology

6.1
Scientific and Engineering Methods

Analysis and synthesis coupling in a loop the theory and the data is the classical method in science and engineering. Descartes united analysis and synthesis in his rules of thought. Vico and Kant, challenging the Cartesian dualism, advanced the idea that knowledge is constructed by the knower. Following similar views, von Uexküll and Piaget demonstrated that knowledge cannot be a passive reflection of reality but has to be an interactive construction.

There are several methodologies describing basically the same scientific and engineering method under different names, for example, interactive man–machine method, iterative two-way approach to the micro–macro link, synthetic microanalysis, modeling relation, goal-directed simulation, elementary loop of functioning, multiagent-based simulations, and so on. These methodologies have similar algorithmic structure based on the main steps: observation, theory formulation, prediction, and testing. In addition, it was observed that strong parallels exist between the structure of scientific and engineering models, the functional organization of organisms as informational systems, and the operational structure of evolutionary devices (Cariani, 1989).

Common to cognition methodologies is the iterative stepwise refinement of the theory, formal or cognitive model, and the constant combination of forward analysis, that is, top-down, with backward synthesis, that is, bottom-up, mediated by real field data obtained from experiment or simulations.

EDOE displays the same basic structure or pattern of scientific and engineering methodologies. The elements of *SKUP* associated with any cognitive methodology are easily identified. The categorical frame may act as the unifying tool for several apparently different methodologies.

However, confronting the complexity barrier, EDOE concentrates on some specific features such as problem representation by several conditioning levels of component designs surrounding the center, the recursivity and focusing methods, the complementarity of forward and backward search, the dissociation and integration techniques, the switch between categorical product and coproduct models, the semiotic

Evolvable Designs of Experiments: Applications for Circuits. Octavian Iordache
Copyright © 2009 WILEY-VCH Verlag GmbH & Co. KGaA, Weinheim
ISBN: 978-3-527-32424-85

closure, and the evolvability potential. Specific features are linked to the wave equation (WE), solution, and interpretation.

6.2
Center Design and Hierarchy

The center design of an EDOE array, at the level $m = 0$, is unconditioned in the sense that it does not depend on other levels. However, an internal hierarchy related to natural order of factors should be taken into account at any level.

Consider the example of the PCB manufacturing process. Four states of the K-center process will be D (design), M (materials), P (processes), and A (applications or tests).

D, M, P, and A have to be considered as the states of the K-process and also as the factors of the central DOE. The manufacturing starts with a selection of these four elements D, M, P, and A. Typically, M depends on D, P depend on M, and so on. All the elements of the engineering design pertain to a variant, say, D. Conditioned on D, one selects the package of materials M. Notation $D > M$ marks out this conditioning. Then the processing (P, $D > M > P$) is selected and finally the testing application (A, $D > M > P > A$) is selected. This hierarchy is suggested by the cyclic character of manufacturing process. The usual choice is in the order D, M, P, A, but any other sorting out is theoretically possible. The manufacturing that is the industrial product realization may start only after the selection of these four elements. It is possible to associate two-digit vectors to the main factors: $D = (1, 1)$, $M = (1, 0)$, $P = (0, 1)$, and $A = (0, 0)$. This choice outlines the above-discussed hierarchy of factors, that is, $D > M > P > A$. This digital form shows that in the background of factor selection the two feature subtexts and a hierarchy are implicit.

Remarkably, the four-factor choice and the linked DOE may be correlated to the Klein 4-group, equivalent to the INRC group studied in genetic epistemology (Inhelder and Piaget, 1958).

Suppose that, at the conditioning level $m = 0$, the initial choice was the sequence $DMPA$. This complete sequence selection allows a comparison with reality following effectively the manufacturing of the product. The selection is correlated to the time frame. After the four-step cycle and the possibility to compare with a real process, the central process starts from fresh. The cyclic structure naturally linked to finite fields or to finite groups justifies the usefulness of these algebraic structures in the study of central processes. D, M, P, and A will be considered as the factors of design of experiment, and concomitantly the same cyclic structure will be outlined by the central DOE.

6.3
Recursivity and Focusing

Recursive modularity, self-similar nesting, or, in other words, the fractal-like natural structure associated with finite fields imposes these basic features on the EDOE construction. The recursivity and the associated redundancy appear as direct consequences of the first-order wave equation and the evolvability request for methodology.

It is accepted that complex problem can be addressed by decomposing them into multiple levels of abstraction. Having several levels of abstraction is better than having a single level. The open problem is: what is the abstraction level to be considered, the number of levels, and how these levels are related to each other and to practice.

Each element of the component DOE is a source of redundancy and a generator of supplementary conditioning levels and DOE.

The primary conditioning level $m = 0$ splits in more sublevels at $m = 1$, then in sub–sub levels at $m = 2$, and so on. Following the confrontation and embedding of reality input, new conditioning levels may be involved in EDOE construction. For the engineering design D, the new splitting is $D1$, $D2$, $D3$, and $D4$. This corresponds to the structure of D. The choice of the design D means the choice of a set of factors $D1$, $D2$, $D3$, and $D4$ with the significance order $D1 > D2 > D3 > D4$. This is the D-process, a component process in PSM or equivalently a component design in EDOE.

The factor M may involve a hierarchy of elements $M1$, $M2$, $M3$, and $M4$, where $M1 > M2 > M3 > M4$. The factor P may involve a hierarchy of elements $P1$, $P2$, $P3$, and $P4$, where $P1 > P2 > P3 > P4$ and so on.

The nature of the EDOE construction implies that the central DOE is variable and may be anywhere.

Any factor or subfactor may be the new center of a basically similar multilevel conditioning structure. Any set of factors may be grouped and considered a new central DOE, for instance, $P1\,P2\,P3\,P4$. The same frame is regenerated recursively. At any moment, however, one center and a hierarchy of levels still exist. The system has to be able to decompose the hierarchy of factors into nested hierarchies of subfactors. Such properties characterize the ultrametric frames (van Rooij, 1978; Schikhof, 1984). Recall that pertaining to the center implies the absence of conditioning. This means that to turn into a new center those factors should lose the previous conditioning. This signifies refocusing on different component processes or designs. Refocusing or regrouping is a method to challenge complexity to ensure evolvability despite finite resources. The same software may run recursively the component DOE from EDOE frame at different levels.

6.4
Problem-Solving Framework for PCB Quality

To illustrate the notion of framework linked to the problem to be solved, the example of the PCB quality will be considered. An industrial product's quality is related to the product factors through complicated nonlinear functions. To determine such functions is at the same time difficult and of limited use. The strategy to face the nonlinearity difficulties is based on the hierarchical structure in "conditioning

Table 6.1 PCB quality framework; conditioning levels.

Design (D)	Materials (M)	Processing (P)	Application (A)
D1-hole size	M1-Cu foil	P1-lamination	A1-optical tests
D2-board thickness	M2-Cu ductility	P2-drilling	A2-electrical tests
D3-innerlayer Cu foil	M3-resin CTE, T_g	P3-conditioning	A3-mechanical tests
D4-layer count	M4-resin/glass	P4-plating	A4-reliability tests
D5-board size	M5-solder mask	P5-imaging	A5-solderability
D6-SMT pitch	M6-surface finish	P6-soldering	A6-chemical tests

$D > M > P > A$; $D1 > D2 > \ldots, > M1 > M2 > \ldots, > P1 > P2 > \ldots, > A1 > A2 > \ldots.$

levels." In technological, biological, or cognitive systems, the configuration of hierarchical representations is a general method for complexity reduction in problem solving. An example demonstrating the main elements of PCB quality is shown in Table 6.1. The hierarchical order is from left to right and in the same column in increasing order of index from top to bottom. In this illustrative case, it is considered that $D > M > P > A$. As the level in the hierarchy of factors associated with the studied problem increases, the degree of stochastic behavior increases too. The timescale necessary for changes decreases in the sense that initial levels in hierarchy, like the design, are slower. Each factor in the quality frame contains secondary factors ranked again in their hierarchical order of significance. The lower is the importance, the greater is the possibility for change. The secondary factors may split in tertiary factors and so on. This gives rise to Table 6.1, containing a detailed and hierarchical description of the PCB quality parameters. Here, CTE denotes the coefficient of thermal expansion; T_g is the glass transition temperature.

The first row in the framework shown in Table 6.1 contains the elements of the central DOE that is D, M, P, and A. It is the level $m = 0$. The columns are the elements of the component DOE ranked in the supposed order of significance. It is the level $m = 1$. The quality framework uses the elements of existing cause and effect diagrams, but the factors should be organized hierarchically and then correlated. The hierarchical ranking is a simple method to approach nonlinearity. The next step, representing what is more specific to EDOE methodology, consists in establishing the branch interactions toward the center that complete the multiple tree construction of hierarchical frame. This represents a major change of conceptual and graphical representation. The nearly linear is replaced or completed by the cyclical; that is, the hierarchic trees are replaced or completed by reticulations or loops. Hierarchically ranged polygonal layers or loops, surrounding various centers, complement the conventional tree frames. It should be emphasized that for EDOE methods the diagram linked to the problem to be solved is not a permanently established thing but an evolvable one. The EDOE framework is modified after each step of analysis to include info concerning the previous measured real data and results. The strategy to perform DOE or failure analysis on this continuously changing space is a self-adaptive stochastic evolution.

6.5
Forward and Backward Search

There are two main directions to be considered for EDOE construction: forward, from center toward outer conditioning levels, and the reverse one, backward, toward inner levels. The two search directions are outlined by the $SKUP$ elements.

The search is based on divergent, forward steps in which alternative concepts are generated according to operators U and on convergent, backward steps in which these alternative concepts are evaluated and selected according to possibilities P.

The transition from forward to backward search is mediated by real field data in S. Forward and backward steps generating a loop are encountered in single-step DOE, and in conventional cybernetics, too. More generally, the so-called scientific and engineering methodology combines analysis and synthesis, decoding and encoding, deduction and induction, divergence and convergence, hypothesis making with experimental verification, and theory with experimentation.

In the EDOE case, the transformations within the global feedback loop and the loop's internal structure, the multiscale loop's interactions, are considered and this proves to be beneficial for accelerated problem solving.

The operators U and the possibilities P are responsible for the decomposition of global loop in a hierarchy of nested loops. The conditions K and the states S steps are involved in iterative cycles. The space K of conditions is generated by the wave equation.

Designed frame steps, in K, are followed by real field evaluations in S and so on. Continuous alternation between the two complementary trends characterizes openended, potentially evolvable systems. It is possible to operate multiple forward steps followed by multiple backward steps.

By the forward type of search, the central design coordinates the evolution at other levels. By the backward type of search, the center summarizes information from different levels. Backward search clusters the entities in classes or assemblies. Forward steps, increasing the conditioning level, are followed by backward steps, decreasing the conditioning level. Specific operators U and possibilities P have to be linked to these transition processes. Following the complementarity between K and S, the complementarity or adjointness between U and P should be examined. Any modification of U induces a modification of P, for instance, through the learning processes that use past experiences.

The forward search is hypothesis driven and often the goal appears as fixed and external to the problem to solve or may be taken into account too early. The reference is to an earlier event to explain the later events. The forward search is a deductive, decoding, and prediction activity.

Observe that running the analysis by the inclusion of more and more conditioning levels maintains the quasilinear character of the model. It is quasilinearity at any level. Globally, a quasilinear, hierarchical frame results. This kind of approach proved to be expensive and inefficient for complexity study. Finally, the designers of experiments were unable to consider all possible parameters and solutions.

The stabilization or decrease of conditioning is mandatory to accommodate the finite nature of the space of search and available time and resources to fix the problem and ensure evolvability. These aspects are linked to the backward search.

Decreasing the conditioning level in outer–inner direction corresponds to data-driven or goal-oriented approach. The backward search is predominantly an inductive, encoding process based on measurement activity.

Backward analysis looks for commonality and similarity, whereas forward analysis focuses on differences. Intermediary backward and forward steps avoid mapping back the experimental information to central control structure too late or too early.

A fast methodology for backward search is to perform classification runs, for rows in DOE matrices, and to compress the data in the backward direction, that is, from outer levels toward the inner ones and the center. For data classification, it is necessary to learn acceptability domains for measurable values and classification thresholds for the similarity degrees in classification. These are heuristics rules based on expertise and physical realizations. For the backward direction to any row, that is, to any experiment, a digit is associated in the DOE matrix at the upper level, more close to the so-called center design.

Some solutions can be grouped together if they are similar, and a set of data is to be replaced by a single symbol. Backward search may consist in replacing old settings by the better performing ones. These replacements may be partial and done in hierarchical order of significance for factors. Some factors may be discarded during the deselection or retrieval operation. Impractical or impossible design solutions are discarded too. The backward step modifies the initial DOE on the basis of experiment and confrontation with real data or constraints. It may consist also in replacing some factors with better performing combination of other factors. The approach from EDOE periphery toward center makes use of the fact that the factors on secondary levels in EDOE are more likely to be modified in accordance with their faster timescales. This accelerates the search and may be critical in ensuring semiotic closure and evolvability (Pattee, 2000).

Obviously, the problem solving is not completed in a single experiment forward and backward. Recursive and iterated search, forward and backward, should be conducted before completing the analysis. EDOE different levels may be activated or deactivated in succession or in parallel. This ensures the EDOE high versatility. It is this continuous alternation between forward and backward search over multiple timescales that allows significant changes in EDOE architecture.

An open problem is to establish the right rhythms of back-and-forth movement along the EDOE frame, the rhythms of different factor modifications, and how to decide when to act. The information criteria, for instance, the entropy and the production of entropy, may be of help to control the moving back and forth among different levels on the hierarchical space of design.

The EDOE center is supposed to have connections to each of the component DOE. The connections are stronger for the immediate outer levels. The fact that any component DOE has its own informational input and output indicates a relative DOE autonomy. It is possible that some information is incorporated in external levels but does not reach the center until the backward search mechanisms are activated. The efficiency of EDOE depends not only on the component DOE structure but also on its connections. In the hierarchy established by using EDOE models, each conditional level shows control features for next levels and noise probabilistic features relative to previous levels. Each level controls more distantly from center levels. This is the forward control model linked to operators U in the $SKUP$. It is relatively easy to model and easily transmissible since it may be stored in restricted memory space. Heritability is associated with the cyclic character captured by the definition of the time T. The associated DOE matrix represents the cyclic memory that should be entirely traversed recursively. It may represent a thorough economy of information allowing the EDOE frame to achieve its function with maximum ease. The complementary backward control model is applied in EDOE frames when the outer level information is compressed or concentrated toward the EDOE center. The possibilities P are linked to backward control model in $SKUP$. This type of control is distributed and in general more complicated than the forward control model. The backward controls are not stored in memory but are part of the real-valued, dynamical rate-depending processes. The dynamical processes are not hereditary transmissible (Pattee, 2000).

Neither control model, forward or backward, has much explanatory and predictive value without the complementary one. However, each control model considered alone can account for a limited level of control.

The controls in EDOE have a hierarchical arrangement, and any of the EDOE levels may occasionally ascend in importance. Disengagement between the center of EDOE and the external levels arrives when these interconnections are interrupted. It is a distorted but potentially useful state of the EDOE network.

It should be observed that the search strategy may be entirely based on the solutions of the first-order wave equation. This offers syntactic advantage and computational efficiency.

6.6
Interactions: Dissociation–Integration

A limitation of quality or reliability standards in different industries is the fact that they do not account for interactions among the components or factors. For instance, the failure rate for an electronic component is considered to be the same for that component regardless of the surrounding components, the substrate, or the process utilized to assemble it into the final electronic product. For nonlinear systems, it is not possible to predict how the system will work by evaluating parts separately and combining them additively.

Similar problem appears in combining specifications, troubleshooting guides, or more generally ontology merging.

An open question for conventional DOE and robust designs is that how they detect and confront nonlinear interactions between parts.

In many conventional designs, the majority of interactions that could possibly contribute to the problem are deliberately excluded.

Conventional DOE takes into account interactions until the second or third order (Montgomery, 1984). For robust design technique, it is considered that the presence of interactions is undesirable (Phadke, 1989). Robust design deemphasizes nonlinear interactions. On the contrary, EDOE methodology seeks interactions and intends to use them. Strong interactions are responsible for complexity, and this in turn imposes and makes possible the EDOE constructivist approach. Interactions may accelerate problem solving since they promote information flow between conditioning levels, that is, between component DOE. Component DOE should include the right expertise in particular directions, but these different kinds of expertise do not mix in the absence of interactions.

Starting EDOE experiments by using Latin squares or Walsh–Hadamard matrices is beneficial since these matrices give reliable estimates of factor effects with fewer experiments compared to the more traditional methods. But in this way the factor interactions may be missed. As expected, some information must be lost when, for example, the 8 runs of a design matrix are used instead of 128 runs of the full factorial design. What is lost is the ability to clearly estimate the interactions between factors and the main effects.

A specific technique to take into account interactions is the dissociation–integration (DI) method discussed in this chapter. For ease, the DI process may be implemented in DOE matrices with fixed number of columns and rows. The DI process means that experiments are performed with parts of more component matrices of experiment designs, lumped together by integration, and then the resulting information is redistributed to the components. The coalesced design may be performed for the whole number of factors of component DOE, or it is possible to select only the main factors from any component design. If it is an interaction between factors, the significance hierarchy for different factors will be modified. Once the coalesced DOE is performed, a redistribution of the factors toward components is completed but this time in the new hierarchical order. The DI process allows a change of information between component DOE. The information resulting from the DI experiment is included in the new hierarchical order of component DOE. Keeping the advantage to work with simple orthogonal or nearly orthogonal experiments, the EDOE may manage the global information via DI experiments. The component DOE represents the visible part in problem solving. The deep levels of DI are not necessarily specified. They embody the hidden information that could be subsequently reactivated. For example, consider that the study starts with a center W4,3,2a matrix of the type 4×3 corresponding to design D, materials M, and processing P. Suppose the factor significance is $D > M > P$. After this first step, the dissociation step is performed for D and M only. Each factor may be decomposed into three subfactors, for example, $D1$, $D2$, $D3$ and $M1$, $M2$, $M3$. The component matrices are again W4,3,2a. The significant subfactors from D and M will be retained, say, $D2$, $D3$ and $M1$, $M2$, and tested in the new DOE center that may now be the one from Table 4.7. The 4×4 matrix from Table 4.7 asks for some "0" settings. They may be assigned to the average or to the best values from the previous experiment. Suppose that $D3$ and $M1$ show significance. The testing restarts with the same 4×3 matrix of the type W4,3,2, in which $D3$ replaces D and $M1$ replaces M.

If strong interaction of subfactors appears, a change of the subfactors hierarchy will result.

The categorical formulation of DI process is correlated to the double interpretation of tensor product in WE model. The categorical product is linked to dissociation or decoding, whereas the coproduct is linked to the integration or encoding trend. The categorical product involves local enlargement of designs while maintaining the old design as the cocoordinating architecture. The coproduct may involve replicating the existing designs and cocoordinating them. Transition from one to another involves and indirectly embeds the information furnished by experimental values. In the case of strong interactions, neither the dissociation nor the integration has much effectiveness for problem solving without the other.

6.7
EDOE Basic Steps

6.7.1
Problem Statement

EDOE construction starts by stating the meaningful question it is expected to answer, that is, with a goal.

6.7.2
Propose the Preliminary Problem-Solving Framework

The potentially primary, secondary, tertiary, and so on factors are enumerated. Experiment objectives, purposes, and limitations of resources, measurable ranges and settings are assigned. The factors associated with the problem and their settings are selected based on the early problem knowledge. At this stage, the preliminary ranking of the factors is proposed. This hierarchy corresponds to the timescale linked to that factor. The closer-to-center levels have longer time horizons but less flexibility.

The timescaling is a significant difference between single DOE strategy and EDOE.

6.7.3
Select the DOE Matrices

This represents the assignment of factors to an array. Among the possible sets of DOE matrices, for n runs, m factors, and s settings, select the Walsh matrix of design Wn, m, s, or Latin square matrices Ln, m, s generated by first-order wave equation and corresponding to required informational entropy or production of entropy.

The center and component DOEs represent the natural units of the problem solving. The number of the first-level components, that is, the first ring, is limited by the number of factors that have been considered in the central DOE. Secondary levels associate DOE matrices with the factors pertaining to the primary level factors and so on in a recursive way.

6.7.4
Run Center Design

The center includes the potentially main contributors to the problem to be solved. The significant factors are identified and confirmed by experiments.

6.7.5
Analyze Results

The calculus is performed for any column of DOE matrices after each experimental step allowing to select the more significant factors. More significant means more correlated. The contribution of factors is evaluated by the correlation, that is, by the spectral analysis of the resulting data. Fourier–Walsh spectrum, or other correlation indices, allows the fast recognition of patterns and factor effects.

6.7.6
Run Multiple Forward and Backward Steps

Perform forward search, that is, DOE tests starting from center toward secondary successive levels, sublevels, and so on following the trajectory of significant factors. The significant factors are identified by the real field experiment and by data analysis for any level.

Each factor of the central DOE may be a source of new DOE at the next level in problem solving.

Backward steps close the loop and generate new DOE in which the previous real data and results are indirectly embedded. Real experiences are accounted for and should be transformed into symbolic descriptions. Classification methods, data compression, and informational calculus are useful for the backward steps.

The decision to run forward or backward and to switch between categorical coproduct and product, the order in which these interweave, should be substantiated by informational criteria and experimental data analysis. By this procedure, the problem formulation or the goal development is included in EDOE construction. The real data are embedded because the search and switch direction at any levels depends on them. This embedding makes the search active and permits evolvability for problem solving.

6.7.7
Perform Dissociation–Integration Experiments

Typically, DI process is between fixed conditioning levels. Two or more DOE may be lumped together in a larger DOE. Then, the factor hierarchy, according to one external real field defined criterion or to informational criteria is reevaluated.

During dissociation–integration processes, "breakthrough" or "resonance" effects may be observed in the sense that once a research direction is fixed, its response may be self-amplified and the solution is clearly highlighted in few steps. DI process modulates these breakthrough solutions.

6.7.8
Establish the New Center Design

The iterative nature of experimentation coupling in succession forward and backward steps is accompanied by increase of problem knowledge.

Forward and backward steps, in succession, DI experiments, offer information for problem-solving continuation.

This allows establishing improved hierarchy, the parameters, and the settings of interest for problem solving. Accordingly, the center is modified.

This step is a problem reformulation and a regeneration of the symbolic description contained by DOE, with possibility of change. It represents another key step to ensure closure and evolvability.

6.7.9
Repeat the Testing Procedure from the New Center Design

The replication of the experiment starting from a new symbolic description as offered by the new central DOE and new matrices of other levels offers potentially efficient search processes.

6.7.10
Run Simulations: Analyze the Solutions of the Problem

This step refers to running different virtual designs or situations and results analysis. The usual procedures ask to build prediction models and be sure the prediction works to modify the supporting software, to establish the conditions for problem solving, improvements, and control to draw conclusions and to evaluate the knowledge gain or loss by applying the EDOE method.

The conclusion of the EDOE refers to failure analysis, corrective actions, optimum or acceptable domains for parameters, and so on. It is necessary to propose appropriate actions and to decide whether another cycle of experiments is needed, to define the significant ways, this means DOE matrices, the array, and frequently used trajectories, allowing problem solving.

6.8
EDOE Frame and *SKUP* Schema

SKUP schema shown in pictorial or diagrammatic forms may be of help to highlight the elements of the cognitive methodology. The EDOE can be visualized by trajectories at different levels, in the space of conditions K, and the space of states S. Moreover, operators U and possibilities P are associated with trajectories illustrating

M22		M23		M32		M33		P22		P23		P32		P33
	M2				**M3**			**P2**				**P3**		
M21		M24		M31		M34		P21		P24		P31		P34
			M							**P**				
M12		M13		M42		M43		P12		P13		P42		P43
	M1				**M4**			**P1**				**P4**		
M11		M14		M41		M44		P11		P14		P41		P44
D22		D23		D32		D33		A22		A23		A32		A33
	D2				**D3**			**A2**				**A3**		
D21		D24		D31		D34		A21		A24		A31		A34
			D							**A**				
D12		D13		D42		D43		A12		A13		A42		A43
	D1				**D4**			**A1**				**A4**		
D11		D14		D41		D44		A11		A14		A41		A44

Figure 6.1 Array of conditions.

the problem-solving history. $SKUP$ schema outlines trajectories rather than the fixed states. The schema are not only spatial frames but also temporal.

EDOE should have an adjustable or mutable construction since it contains interacting components DOE subjected to continuous reorganization after confronting the reality.

Figure 6.1 consists of a complete three-level array of conditions.

Here, the notations are standard, that is, D-design, M-materials, P-processes, A-application corresponding to the level $m = 0$. The conditions were represented as elements of a cyclic loop. $D1$, $D2$, $D3$, $D4$ are the factors of D corresponding to the level $m = 1$. Then, $D11$, $D12$, $D13$, $D14$ are the subfactors of $D1$ and correspond to the level $m = 2$. They were represented as elements of a cyclic loop too. Figure 6.1 contains the semiotic network of all the possible conditions K, that is, the selected factors, to be grouped in DOE. It should be emphasized that this semiotic network represents particular solutions of the wave equation.

One outcome of the complexity is that currently the designer cannot adequately explore all the design alternatives and select the best.

Consequently, $SKUP$ schema may include only some of the conditions K and the corresponding states S also. Consider only a fragment of the Figure 6.1. For three-level evolution, $m = 0$, $m = 1$, $m = 2$, the $SKUP$ consists of the vectors $S = (s^0, s^1, s^2)$, $K = (k^0, k^1, k^2)$, $U = (u^0, u^1, u^2)$, and $P = (p^0, p^1, p^2)$.

Figure 6.2 illustrates the *SKUP* schema, with conditions and states for two levels, $m = 0$ and $m = 1$, only. D, M, P, and A are the conditions at the level $m = 0$. Let $D = k_0^0$, $m = k_1^0$, $P = k_2^0$, and $A = k_3^0$. The upper index refers to the level while the lower index refers to the time step. It should be emphasized that despite notation the time steps at different levels may be different, and these multiple scales are key features for evolvability. The states and conditions at the level $m = 0$ are represented by high-thickness border cells (Figure 6.2).

The system initial state is s_0^0. With possibility $p^0(k_0^0|s_0^0)$, the condition k_0^0 is selected. This condition is a digit symbolizing a specific engineering design D. This may be a matrix corresponding to D-DOE. Based on this, the operator $s_1^0 = u^0(k_0^0, s_0^0)$ allows the transition to the new state s_1^0. This state is the realization of the design. Then, with possibility $p^0(k_1^0|s_1^0)$ the new condition k_1^0 arises. This condition symbolized by a digit corresponds to the selection of materials M. In the new condition, the operator $u^0(k_1^0, s_1^0) = s_2^0$ allows the system to reach the state s_2^0. This corresponds to the completion of design and materials.

Observe that $s_1^0 = u^0(k_0^0, s_0^0)$ implies $s_2^0 = u^0(k_1^0, u^0(k_0^0, s_0^0))$.

With possibility $p^0(k_2^0|s_2^0)$ the process k_2^0, that is, P, is selected and finally the new state $s_3^0 = u^0(k_2^0, s_2^0)$ results. This may correspond to the processed product. It represents the succession of realized design, materials, and processes.

Observe that $s_3^0 = u^0(k_2^0, u^0(k_1^0, u^0(k_0^0, s_0^0)))$.

The product will be tested at the level $m = 0$ in the condition A denoted by k_3^0. After test the state is $s_4^0 = u^0(k_3^0, s_3^0) = u^0(k_3^0, u^0(k_2^0, u^0(k_1^0, u^0(k_0^0, s_0^0))))$.

The states result not necessarily in a recursive way since in practical cases the operators may vary with the step.

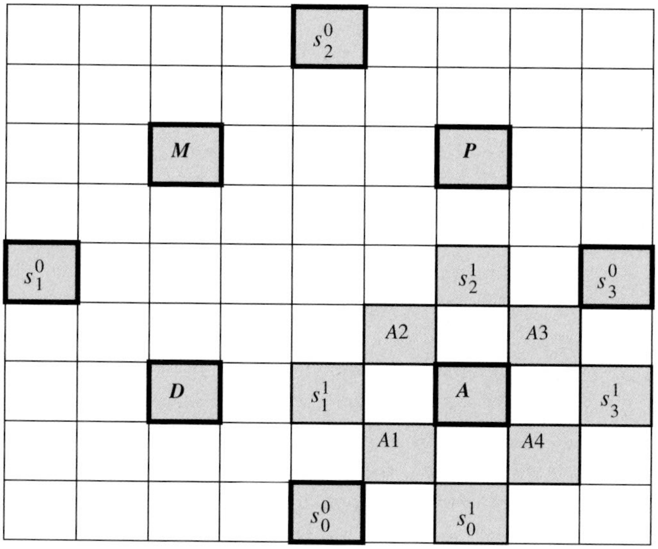

Figure 6.2 Two-level *SKUP* schema ($m = 0$, $m = 1$).

The states at the level $m = 0$ are $s_0^0, s_1^0, s_2^0, s_3^0, s_4^0$. The interpretation of the high-thickness border cells trajectory is as follows: from the state s_0^0 through condition k_0^0 toward the state s_1^0, then through condition k_1^0 toward the state s_2^0, and so on.

Piaget (1971) uses to represent this kind of schema by cycles as

$$\left(s_0^0, k_0^0\right) \to \left(s_1^0, k_1^0\right) \to \left(s_2^0, k_2^0\right) \to \left(s_3^0, k_3^0\right) \to \left(s_0^0, k_0^0\right).$$

The relation with Kant schema as a collection of actions and thoughts, allowing to interact with the world and to solve problems, and with von Uexküll functional cycle is obvious. Correspondence with the categorical frames is of interest too. The Piaget concept of assimilation may be linked to the categorical limit. The concept of accommodation refers to the formation of new schema, for instance, by inclusion of other pairs (states, conditions) and may be linked to the categorical colimit. Equilibration ensures the *SKUP* schema commutativity (Piaget, 1990).

Moreover, the perception and action of von Uexküll functional cycle can be viewed as adjoint functors. The functor U, linked to categorical limit, is the adjoint of the functor P, linked to categorical colimit.

If the experiment analysis shows that the factor A is the more significant factor, the analysis may be continued at the level $m = 1$ for different test conditions $A1 = k_0^1$, $A2 = k_1^1$, $A3 = k_2^1$, and $A4 = k_3^1$. This means to perform four different tests.

The states and the conditions at the level $m = 1$ are represented in Figure 6.2 by medium-thickness border cells. The system initial state at the level $m = 1$ is s_0^1. With possibility $p^1(k_0^1|s_0^1)$ the condition k_0^1 arises. The condition is a digit symbolizing a specific test. Based on this, the operator $u^1(k_0^1, s_0^1) = s_1^1$ describes the transition to the new states_1^1. Then with possibility $p^1(k_1^1|s_1^1)$ the new condition k_1^1 arises. In the new condition, the operator $u^1(k_1^1, s_1^1) = s_2^1$ allows the system to reach the state s_2^1.

Observe that $s_2^1 = u^1(k_1^1, u^1(k_0^1, s_0^1))$ and $s_3^1 = u^1(k_2^1, u^1(k_1^1, u^1(k_0^1, s_0^1)))$.

The states at the level $m = 1$ are s_0^1, s_1^1, s_2^1, and s_3^1.

The condition at the level $m = 1$ is represented by the loop $A1A2A3A4$, that is, k_0^1, k_1^1, k_2^1, and k_3^1.

The interpretation of the medium-thickness border cells trajectory is as follows: from the initial state s_0^1 through condition k_0^1 to the state s_1^1, then through condition k_1^1 to the state s_2^1, and so on.

Due to presentation restrictions, Figure 6.2 typically illustrates only two successive levels, in this case $m = 0$ and $m = 1$. Suppose that a more detailed study is necessary. The same frame may be used to outline levels $m = 1$ and $m = 2$.

A figure similar to Figure 6.2 may be useful but the level $m = 0$ is discarded and replaced by the level $m = 1$.

Figure 6.3 shows the states and the conditions for next two levels.

The trajectory at the level $m = 1$ is represented by medium-thickness border cells and that at the level $m = 2$ is represented by standard-thickness border cells.

In this example, it was established that the condition $A4$ needs a more detailed study. The new conditions are $A41 = k_0^2$, $A42 = k_1^2$, $A43 = k_2^2$, and $A44 = k_3^2$. The states at the level $m = 2$ are s_0^2, s_1^2, s_2^2, and s_3^2.

Figure 6.4 shows an example of three-level *SKUP* schema.

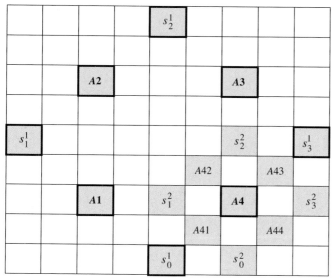

Figure 6.3 Two-level SKUP schema ($m = 1$, $m = 2$).

The trajectories are illustrated by high-thickness border cells for $m = 0$, by medium-thickness border cells for $m = 1$, and by standard-thickness border cells for $m = 2$.

In this case, it is supposed that A is the significant factor at the level $m = 0$, $A4$ is the significant factor at the level $m = 1$.

This imposes the experiment at the level $m = 2$.

The interpretation of the standard-thickness border cells trajectory is as follows: from the state s_0^2 through condition k_0^2 to the state s_1^2, then through condition k_1^2 to the state s_2^2, and so on.

Observe that

$$s_1^2 = u^2\left(k_0^2, s_0^2\right), \quad s_2^2 = u^2\left(k_1^2, u^2\left(k_0^2, s_0^2\right)\right) \text{ and } s_3^2 = u^2\left(k_2^2, u^2\left(k_1^2, u^2\left(k_0^2, s_0^2\right)\right)\right).$$

The recursive character of EDOE is evident from Figures 6.2–6.4.

The high-thickness border cell loops overlap the smaller medium-thickness border cell loops and these in turn overlap the smaller, standard-thickness border cell loops.

It is a large loop with superposed smaller loops and so on.

Thus, EDOE may result from self-replicating loops with universal construction and universal computation.

Observe that generally $K = ((k_0^0, k_1^0, \dots), (k_0^0, k_1^1, \dots))$, $S = ((s_0^0, s_1^0, \dots), (s_0^1 s_1^1, \dots))$, $U = (u^0, u^1, u^2, \dots,)$, and $P = (p^0, p^1, p^2, \dots,)$..

Summing up, K is the symbolic description of the system, and in this case, the planned choices for factors such as design, materials, processes, and applications or tests at different conditioning levels. K elements are linked to matrices in DOE and are generated by the wave equation.

The states S represent the realization of those plans or experiments. The states S are real, not symbolic. They correspond to material functions, for example, plating,

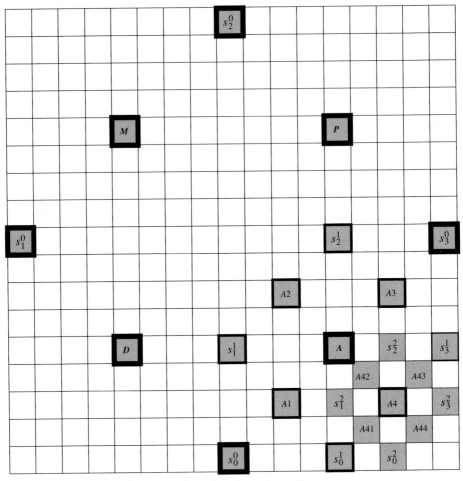

Figure 6.4 Three-level *SKUP* schema ($m = 0$, $m = 1$, $m = 2$).

soldering, and so on. The S processes are described by dynamical models. It is possible to make only partial use of states. For instance, DOE may limit the interest to experimental results only after a complete loop DMPA, not before. This choice corresponds to testing the final product only.

The operator U characterizes the capability to pass from plans or symbols to reality based on previous reality and on new plans. In one of the simplest cases, U describes the trajectories in real-state space by concatenation of successively realized plans.

The states' capability to reinitiate the plans and modify the symbolic description is characterized by the possibility P. P describes the selective activation or deactivation of conditions pertaining to K.

Figure 6.4 outlines the sparseness of functional sequences when compared to the size of the search space as shown in Figure 6.1.

Due to the size of search space, EDOE is confronted with the apparent improbability of chance to produce any successful solution of the problem. But in fact the problem solving is not a blind search. The wave equation allows modifying the searchable domain and the search velocity by adding more levels and scales to the search process. Moreover, interaction with real data may accelerate discovering solutions.

The loop at the level $m = 0$ is completed by the small loops at the level $m = 1$ and by the smaller loops at $m = 2$.

6.9
Comparison of EDOE with other Methods

Well-known graphical representations of problem-solving methods are "quasilinear" or hierarchical, flow charts, reliability trees, cause and effect diagrams. Often, these "quasilinear" representations and the corresponding analytical methods prove to be inconclusive and have to be completed and developed by a new methodology. One of the main differences of EDOE from conventional methods is the hybrid, that is, cyclic and tree-like, character of EDOE resulting from the nonlinear interactions of the factors. Due to these interactions, EDOE may close the frames of factors by changing the linear hierarchy illustrated by fishbone or hierarchic trees by reticulated trees and multilevel networks around variable centers.

EDOE switches between cyclic and hierarchical trees and organizes the space of test conditions. This is linked to the switch between categorical coproduct and product and represents a significant capability for complex problem solving.

EDOE appears as cognitive architectures, cooperative networks of active, and interdependent but partially autonomous component DOE, developed to perform and represent problem solving. This results in the structured EDOE built by hierarchical and recursive concentric levels or sublevels of DOE.

A feature that makes EDOE more efficient is the way it bypasses single-level capability and jumps to multiresult and multicontext schema.

The transition from a matrix design to other ones needs and embeds the real field-measured data input.

When a problem to solve is exposed to the EDOE architecture, it will create activation patterns at multiple levels, which are actually different layers of organization.

EDOE integrates hierarchy and interactions, linearity and circularity, and sequential and parallel approach. The levels and the sublevels are linked to each other and normally have input toward centers, according to the possibilities P, and receive input from the center and environment, according to the operators U.

The net-like and recursive appearance of the EDOE is imposed by the factor's nonlinear interactions and by the need to solve the problem at the right time. Instead of being considered as an obstacle, the strong interactions that induced the specific multilevel frame may simplify and accelerate the search. In EDOE, the digital DOE steps, in K, alternate with measurements and evaluation steps for S, followed by DOE matrix reorganization. Real field measurements and DOE digital frames have been

Table 6.2 Comparison of different DOE methods.

Characteristics	Conventional DOE, robust design	EDOE
Resources	Considered unlimited or very large	Definite resources, time frames
Objectives	One solution, one goal, monocentric, search for ideal optimal solutions	General goals, multicentric, workable, acceptable, viable solutions
Strategies	Prescribed, fixed	Adaptive, evolvable
Matrices	Fixed matrices – specific, developed Defined-based on previous learning Conventional design-detailed models	Constructed, self-adaptive Sensitive to case study, undefined Generated by equation and evolved
Methods	Single DOE Sequential	Frameworks of small DOE Parallel and sequential
Required data	Clear steps, complete data	Incomplete data and variables
Factor hierarchy	Ignored; Pareto charts; inner–outer factors	Heterarchy, evolvable frameworks
Approach	Mainly deductive Isolate from medium	Deductive and inductive Builds, dissociates, and reconstruct
Interactions	Typically neglected, linear or quasilinear	Considered from start, useful
New experiments	Controlled use, confirmation experiment	Self-adaptive, evolution, minimal
Selectivity	Selective, robust to noisy	Robust to redundant info and noisy
Criteria	Average, variance, signal/noise	Informational entropy, distance
Calculus	Real field	Real and other than real fields
Data analysis	Developed, completed	Adaptive, interactive, online
Learning	Early learning	Beyond learning, creative
Time concept	Neglects time or considers linear time only	Timescales, cyclic time frames
Complexity	Considered as enemy, avoided	Considered as ally, accepted

viewed in their complementarity. The strategy is to use real field data to determine whether or not a process needs to be improved, to use EDOE to improve the process, and to measure the results of EDOE by continuing to collect real field data.

Some distinctive features for EDOE are summarized in Table 6.2.

Differences between EDOE and most conventional problem-solving methods are rooted in the fact that EDOE is not preprogrammed but actively builds its own structure.

EDOE is not designed for any particular problem, but it is a general cognitive architecture that can be applied to different domains and problems. The objective is not a passive reflection of reality but an active construction of an experiment or a device.

EDOE methods start with model generation of the most informative DOE matrices. The DOE matrices differ from those used in traditional DOE or robust design with few exceptions. EDOE matrices are formulated to offer maximum information. EDOE methods may use both types of factor analyses, that is, columns in DOE matrix as well as the classification methods for runs, that is, rows in DOE matrix. Classification procedures accord different weights to contributors to the problem to solve. This is a way to include the reality input in the EDOE frame.

The factor interactions are acknowledged and systematically exploited as opportunities in the EDOE method. Some component DOE or some interconnections can be intermittently inactivated, permitting in this way innovative approach and less time to solve the problem. The switching mechanism is of help in this case. EDOE emphasizes the importance of real context. The involved conditioning levels highlight specific contexts. EDOE embeds real field results in the selection of the component DOE and of the associated factors and matrices for any DOE. Constructing step by step the matrix arrays, with measurements and analysis between DOE steps, it is possible to account for the difficulties in ensuring the settings for factors, the limitations in conducting experiments, the availability of resources, and the existing opportunities. EDOE methodology makes use of the incomplete information that the rejects or the errors contain. EDOE is working with orthogonal or non-orthogonal experiments, with incomplete data and looks for compromise solutions.

The standard DOE methods and robust design methods have deliberately selected features and proved their efficiency in several acknowledged situations. Differences in the methods come about because companies that were concerned with specific practical problems and contexts developed different strategies.

Following the advent of robust design, that is, Taguchi methods for quality, there has been a debate on differences between various points of views on experiment design. Differences between conventional "Western" designs and Taguchi "Eastern" approach to quality have been examined (Lochner and Matar, 1990). During this debate, it was established that the design matrices used by Taguchi have been known and applied for many years as mathematical tools. In fact, robust design studies make heavy use of known orthogonal arrays, with alterations that correspond to specific and challenging choices (Dey, 1985; Mori, 1995; Hedayat, Sloane and Stufken, 1999). It was acknowledged that in practice we need to use several types of DOE.

The frame imposed by conventional DOE and by some robust designs is fixed from start, without memory of the values of factors during experiment. It may happen that ideal experiments as recommended by the conventional DOE are impossible to achieve in practice or that the resulting best settings do not correspond to one of the performed experiments in the DOE fixed matrix. The standard Taguchi designs are linked mainly to the coproduct interpretation of tensor product in the WE solution. Such designs follow the coproduct "∪" way in categorical frame. This allows screening of the number of factors in components DOE but increases the risk of missing significant ones. To compensate this drawback of orthogonal design, a procedure of dealing with uncontrollable or outer design factors was developed (Taguchi, 1986). Robust design emphasized the useful distinction between control and error factors. These are in fact enclosed by the focus on center and secondary levels in the EDOE approach. Notice that EDOE does not restrict to two scales or levels only. For dispersion experiments, Taguchi used the so-called product designs in which control factor settings are given by one orthogonal array, error factor setting are given by another orthogonal array, and for each combination of control factor and error factor an observation is taken. In this case, both categorical product "×" and categorical coproduct "∪" are involved. Close to this idea, the compound orthogonal designs share the attractive properties of Taguchi's designs, but they are often smaller for given strength or stronger for given size (Rosenbaum, 1996; Hedayat and Stufken, 1999).

Some conventional statistical designs follow preponderantly the product "×" way in categorical tensor interpretation. In such cases, the number of solutions will increase as the analysis gets increasingly detailed.

Note that the EDOE develops a strategy to involve in significant order both the categorical product and the coproduct of component designs to construct the space K of EDOE. The switch between the two types of WE solutions is interceded and determined by real field data. The rhythm in which the categorical product and coproduct steps are performed may be critical for complex problem solving.

The use of traditional statistical DOE, of robust design, or of any other modeling methods depends on the windows of application. Robust design methods might have an advantage in the speed and cost of industrial innovation in the mechanical segments of the industry but little advantage in other domains, notably, chemical, biochemical, and electronic industries (Mansfield, 1988). Robust design methods seem to be more pertinent to traditional manufacturing environment. For high complexity problems, conventional statistical designs are not the recommended tools due to the excessive expansion of the search space.

Being able to balance the increase in the effectiveness of exploration with minimum loss of the richness of the solution space explored, the EDOE constructivist methods are essential for complexity barrier breaking.

Part Three
Case Studies

Evolvable Designs of Experiments: Applications for Circuits. Octavian Iordache
Copyright © 2009 WILEY-VCH Verlag GmbH & Co. KGaA, Weinheim
ISBN: 978-3-527-32424-8

7
Solder Wicking

7.1
Illustrative Failure Analysis

To illustrate EDOE methodology, the failure analysis for solder wicking failure on flip-chip is presented in what follows next (Iordache and Hu, 2000a, 2000b). Flip-chips have been used in a variety of electronic systems, including computers, telecommunications, and automotive electronics. Solder wicking phenomenon appears when the solder moves toward preferred areas of the leads or pads. Figure 7.1 shows the metallic pads and the solder penetrating the space between the foil and the pad. Note that in electronic microscopy photos the dark areas correspond to low atomic numbers such as in organic compounds, whereas the bright areas correspond to high atomic numbers such as in metals.

The solder eutectic shows tin-rich gray area and lead-rich white area.

The hierarchical structure of factors, similar to that from "cause and effect" diagrams is maintained here, but only as a starting point. The center in the EDOE linked to this failure includes the following three factors, the design $D = F1$, the materials $M = F2$, and the processing $P = F3$. The EDOE primary elements are ranked from left to right in the proposed order of possible contribution to failure. To start with, it is considered that the design D is more important for failure analysis than materials M, and these are more important than the processing P. More important means more correlated. Materials M are conditioned by the design D, and processes P are conditioned by materials M and by the design D. In abbreviated form the same can be represented as $D > M > P$ or $F1 > F2 > F3$. For the wicking examples at the secondary structure stage, the design factors are, for example, the pad design $F11$, the leads design $F12$, the grid size $F13$, and so on. Suppose that the significance order for design factors is $F11 > F12 > F13 \ldots$. The material factors are denoted by $F21$, $F22$, $F23$, and so on. Consider that the importance order is $F21 > F22 > F23$. The analysis starts with the center of the EDOE. From this, the $F1$, $F2$, and $F3$ factor trajectories diverge.

Suppose that there exist three settings, that is, three types of designs $D = F1$, denoted by $\{-1, 0, 1\}$, three classes of materials $M = F2$, and three processing lines $P = F3$. In the considered case, three different types of samples (old, new, and

Evolvable Designs of Experiments: Applications for Circuits. Octavian Iordache
Copyright © 2009 WILEY-VCH Verlag GmbH & Co. KGaA, Weinheim
ISBN: 978-3-527-32424-8

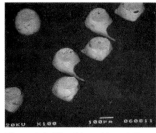

Figure 7.1 Flip-chip solder wicking.

modified) have been designed and manufactured in three different facilities (denoted by KL, GB, and PC).

Table 7.1 shows the factors.

It is possible to simulate wicking condition by depth of the flip-chip sample in solder at 245 °C for 25 s. For less than 25 s deep time, no wicking was observed.

The interest was twofold. First, if the failure is present, to establish the failure reason, and second, if the failure is absent, to establish the corrective actions.

The proposed DOE matrix and the corresponding results are shown in Table 7.2 (see Table 4.12).

The spectral analysis is performed. It results in $q_1 = -0.1$, $q_2 = 1$, $q_3 = 0.3$.

The more correlated to results factor is $F2$, followed by $F3$ and $F1$.

Table 7.1 Factors for central DOE.

Factor	"0"	"1"	"−1"
D-$F1$	Old	New	Modified
M-$F2$	KL	GB	PC
P-$F3$	KL	GB	PC

Table 7.2 Center matrix C.

D-$F1$	M-$F2$	P-$F3$	Fail (%)
−1	−1	−1	0.2
0	0	0	0.1
1	1	1	0.6
−1	0	1	0.3
0	1	−1	0.5
1	−1	0	0.1
−1	1	0	0.4
0	−1	1	0.2
1	0	−1	0.1

Table 7.3 Factors for *M*-materials.

Factor	"−1"	"1"
F21	Full gold	Selective
F22	Supplier *S/M-D*	Supplier *S/M-C*

The right order for factors of the center design is $F2 > F3 > F1$. In this stage of the analysis, one ignores exactly which subfactors of the materials are responsible for the failure.

Since $M = F2$ was significant, supplementary analysis will be performed in the $F2$ direction. This is a forward step search. The worst condition results have been considered, that is, $F1 = 1$, $F2 = 1$, $F3 = 1$, and the factor $F2$ and then $F3$ have been modified. The plant-related materials factors, such as surface type under mask type $F21$, as well as the solder mask type itself $F22$, will be involved. Table 7.3 contains the meaning of factors for the next level of DOE. There exist two types of surface finish $F21$. Here "−1" denotes the full gold, and "1" denotes the selective gold. There exist two classes of solder mask $F22 = \{-1, 1\}$, where "−1" denotes the solder mask supplier D and "1" denotes the solder mask supplier C.

The matrix for this DOE and the corresponding results has been included in Table 7.4.

This matrix is based on solutions of the first-order wave equation wave equation (WE) in GF(2).

It is the convection model solution at $T = 0$ and $T = 1$ followed by pasting down of results.

Again "0" was replaced by "−1".

The associated spectrum is $q_{21} = -0.5$, $q_{22} = -0.1$. Therefore, $F21 > F22$, that is, $F21$ is more correlated to failure. The DOE shows that the failure is associated with "full gold" technology and to the supplier "$S/M-D$". An immediate corrective action would be to use "selective gold" technology and the supplier "$S/M-C$".

The next search for failure is for "full gold" and supplier D for mask. This means that $F21 = -1$ and $F22 = -1$, which corresponds to the worst conditions for materials.

The processing subfactors are $F31$ and $F32$. The physical meaning is as follows: $F31$ the time for cure and $F32$ the pressure in developer. Here, "1" denotes the "new" setting of the parameters. The "old" values of the parameters "−1" denote the out-of-control process.

Table 7.4 Matrix for *M*-materials.

F21	F22	Fail (%)
1	1	0.3
−1	−1	0.6
1	−1	0.3
−1	1	0.5

Table 7.5 Factors for P-processes.

Factor	"−1"	"1"
F31	60 min	40 min
F32	60 psi	45 psi

1 psi = 6.9 kPa.

Table 7.5 contains the factor settings.

The results have been shown in Table 7.6.

The spectrum is $q_{31} = -0.5$, $q_{32} = -0.1$.

Data analysis shows that the correlation or the significance order for factors is $F31 > F32$.

The failure is associated with time for cure 60 min and pressure 60 psi (1 psi = 6.9 kPa).

Finally, a change of the center design is performed. To take into account that F2 and F3 are more significant, only these will be modified.

Restricting modifications to F2 and F3 may accelerate the problem solving since typically a change of F1 is slower. F2′ replacing "1" by "1′" in the column F2 of Table 7.2 signifies that $F21 = -1$, $F22 = -1$ and F3′ replacing "1" by "1′" in the column F3 of the Table 7.2 signifies that $F31 = -1$ and $F32 = -1$ (Table 7.7).

Table 7.6 Matrix for P-processes.

F31	F32	Fail (%)
1	1	0.3
−1	−1	0.6
1	−1	0.3
−1	1	0.5

Table 7.7 Center C′.

F1	F2′	F3′	Fail (%)
−1	−1	−1	0.2
0	0	0	0.1
1	1′	1′	0.7
−1	0	1′	0.5
0	1′	−1	0.7
1	−1	0	0.1
−1	1′	0	0.5
0	−1	1′	0.5
1	0	−1	0.1

The new center matrix includes the significant factors of the previous center C and of the significant component design performed. This resulted by a DI process. Observe that for this analysis, the matrix of the center C' was selected to be formally the same as for the center C.

7.2
Illustrative EDOE Frame

Figure 7.2 shows an illustrative EDOE frame for solder wicking failure analysis. The center contains three factors D, M, and P. Component DOEs have been performed for materials M and processes P only.

The indicated results of the row experiment are percentage of failures.

The EDOE is useful not only to establish failure root, but it also has predictive capabilities. This capability illustrates the backward search step, subsequent to the forward search. This is specific to EDOE methodology and is related to the possibilities P in the associated $SKUP$ quadruple.

The method for backward search consists in performing classification runs, for rows in DOE matrices, and compressing the data in the backward direction, that is, from outer conditioning levels toward the inner ones and the center. For data classification, it is necessary to learn acceptability domains for measurable and

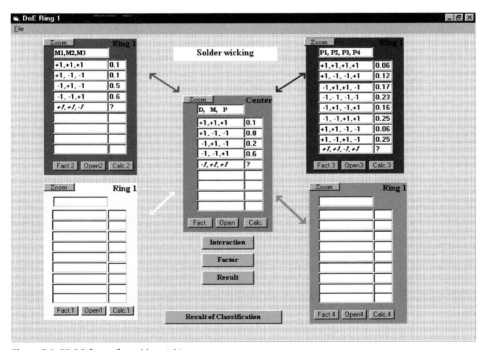

Figure 7.2 EDOE frame for solder wicking.

classification thresholds for the similarity degrees in classification. For the backward direction to any row, that is, to any experiment, a digit is assigned in the DOE matrix at the next level, more close to the center design.

New vectors, that is, settings of parameters for an experiment may enter as new rows in the DOE matrices. New vectors have been denoted as italicized rows.

Suppose that the processing vector is $\langle 1,1,-1,1 \rangle$. This is the last vector in the DOE component P. This nonperformed experiment will be classified as the experiment 1 or 7 at the acceptable similarity level $T = 0.75$. This allows predicting that a 6% failure percentage will result in this case. This classification step emphasizes the importance of ranking the factors according to their right contribution to failure. To any new settings of parameters, acceptable degrees of similarity thresholds T are associated. They depend on accumulated practical expertise and the user's feedback. Each run is classified in the high-quality group, say "1" or in the low quality group, say "−1," and the digit is shifted in the backward direction toward the center. If the failure percentage 10% is considered as a good result, the vector M: $\langle 1,\ 1,\ -1 \rangle$ is shifted backward as a "1" in the column M of the center. Coupled with the P vector $\langle 1,1,-1,1 \rangle$ offering another "1" in the center matrix and with the "new design," say "−1," it results in the center matrix the vector $\langle -1,\ 1,\ 1 \rangle$. This is classified by similarity with the fourth run in the center. This allows predicting a 20% of failures. The conclusion is that the "new design" is the reason of failure increase.

7.3
SKUP Schema for Solder Wicking

Figure 7.3 illustrates some elements of the *SKUP* schema linked to the failure analysis. It contains conditions, states, and trajectories.

The notations are D for engineering design, M for materials, P for processes, and A for tests or applications at the level $m = 0$. The notations are M1 and M2 for the factors of M and P1 and P2 for the factors of P at the level $m = 1$.

The *SKUP* contains the elements $S = (s^0, s^1)$; $K = (k^0, k^1)$; $U = (u^0, u^1)$; $P = (p^0, p^1)$. K is the symbolic description of the selected factors in DOE.

S represents the real product following each achievement of the symbolic steps in K. Any experiment corresponds to a new trajectory in the *SKUP* schema.

Let $D = k_0^0$, $M = k_1^0$, $P = k_2^0$, and $A = k_3^0$. The trajectory at the level $m = 0$ is represented by high-thickness border cells.

The initial state of the system is s_0^0. With possibility $p^0(k_0^0|s_0^0)$, the condition k_0^0 is selected.

This is the digit symbolizing the specified design D. On the basis of this, the operator $u^0(k_0^0, s_0^0) = s_1^0$ defines the transition to the new state s_1^0. Then with possibility $p^0(k_1^0|s_1^0)$, the new condition k_1^0 arises. This condition is symbolized by a digit corresponding to the selection of materials M. According to the operator $u^0(k_1^0, s_1^0) = s_2^0$, the system reaches the state s_2^0. With possibility $p^0(k_2^0|s_2^0)$, the processes $k_2^0 = P$ are selected and finally the product $u^0(k_2^0, s_2^0) = s_3^0$ results. It will be tested at the level $m = 0$ in the condition $A = k_3^0$.

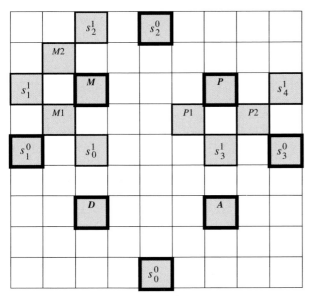

Figure 7.3 *SKUP* schema for solder wicking.

After the test, the state is s_4^0. This may be compared with s_0^0.

Observe that $s_1^0 = u^0(k_0^0, s_0^0)$, $s_2^0 = u^0(k_1^0, u^0(k_0^0, s_0^0))$, $s_3^0 = u^0(k_2^0, u^0(k_1^0, u^0(k_0^0, s_0^0)))$, and $s_4^0 = u^0(k_3^0, u^0(k_2^0, u^0(k_1^0, u^0(k_0^0, s_0^0))))$.

The initial state and the conditions appear to be the determinant for the trajectory.

The interpretation of the high-thickness border cells trajectory is as follows: from the state s_0^0 through condition k_0^0 toward the state s_1^0, through condition k_1^0 toward the state s_2^0, and so on. Cycles as $(s_0^0, k_0^0) \rightarrow (s_1^0, k_1^0) \rightarrow (s_2^0, k_2^0) \rightarrow (s_3^0, k_3^0) \rightarrow (s_0^0, k_0^0)$ correspond to the schema.

Since several experiments show that M and P are significant factors, the analysis may be continued at the level $m = 1$ for the conditions $M1 = k_0^1$, $M2 = k_1^1$, $P1 = k_2^1$, and $P2 = k_3^1$. The trajectory at the level $m = 1$ is shown by medium-thickness border cells. The system initial state at the level $m = 1$ is s_0^1. With possibility $p^1(k_0^1|s_0^1)$, the condition k_0^1 arises. This is a digit symbolizing $M1$. On the basis of this, the operator $u^1(k_0^1, s_0^1) = s_1^1$ describes the transition to the new state s_1^1 and so on. With possibility $p^1(k_1^1|s_1^1)$, the condition k_1^1 arises. This is a digit symbolizing $M2$. On the basis of this, the operator $u^1(k_1^1, s_1^1) = s_2^1$ describes the transition to the new state s_2^1 and so on. Observe that the high-thickness border cells trajectory at the level $m = 0$ is complete, while the medium-thickness border cells trajectories are incomplete due to the fact that experiments allow to reduce the number of tested conditions. The coexistence of complete and incomplete loops is an EDOE technique to accelerate problem solving.

8
Reliability Analysis

8.1
EDOE for Reliability

The main way to characterize printed circuits' reliability class is through the market. From this point of view, the consumer, the industrial, and the military aerospace markets appear to be the main classes. Manufacturers of consumer electronics operate by considering that high quality ensures high reliability. Manufacturers of reliable consumer products implement SPC and most operate without any outside standard except customer feedback. The industrial market activity is based on standards. The industrial product is expected to pass a defined set of qualification tests. Reliability methods for military aerospace market are well defined by the customer in standards that establish not only the required reliability level but also how the product must be manufactured, tested, and screened.

The guidelines for reliability testing contains information about environmental conditions for electronics used in consumer products, computers, telecommunications equipment, commercial aircraft, automobiles, military ground and ship applications, space equipment, and military avionics equipment. The objective of reliability tests is to determine the PCB construction and manufacturing giving acceptable life test for a specified reliability test and application. For many years, the PCB reliability studies have been focused on plated through hole (PTH) performances. The elements contributing to PTH reliability may be ranked in hierarchical order as given in Table 8.1.

Online control of the reliability can be performed by accelerated stress tests (ASTs) such as solder shocks (S/Ss), liquid–liquid (L/L), interconnect stress testing (IST), and so on.

IST is one of the widely used accelerated reliability tests. In IST, the daisy chains of copper barrels and interconnecting tracks are heated by passing DC current. The heating temperature is below 250 °C, the cooling is 25 °C, the dwell time is approximately 3 min, and the rejection criterion is at 2–100% of the electrical resistance. Usually, the heating temperature is 150 or 200 °C, and the reject criteria experience a 10% increase. The number of thermal cycles to failure and the electrical resistance are recorded.

Evolvable Designs of Experiments: Applications for Circuits. Octavian Iordache
Copyright © 2009 WILEY-VCH Verlag GmbH & Co. KGaA, Weinheim
ISBN: 978-3-527-32424-8

Table 8.1 PTH reliability framework.

Design, D	Materials, M	Process, P	Application, A
D1-Hole size	M1-Resin T_g	P1-Cu plate thick	A1-Assembly
D2-Board thick	M2-Cu elongation	P2-Drilling quality	A2-Solder shocks
D3-Pad size	M3-Z-Expansion	P3-Surface finish	A3-Liquid–liquid
D4-Cu foil type	M4-Delamination time	P4-Cu thick CV	A4-IST
D5-Number of layers	M5-T_g CV	P5-Cu elongation	A5-Air–Air 1000
D6-Grid size			A6-Field

T_g, glass transition temperature; CV, coefficient of variation. $D > M > P > A \ldots$; $D1 > D2 > D3 >, \ldots,$ $M1 > M2 > M3 >, \ldots, P1 > P2 > P3 >, \ldots, A1 > A2 > A3 > \ldots$.

The tests may be performed on boards or done on specially designed coupons.

Developed minicoupons represent embedded DOE including variability of the most critical elements of the PCB (Iordache and St. Martin, 1997). They are designed to embed matrices of Walsh type *W* or Latin squares *L*. The minicoupons are implanted on the frame of production panels.

This allows reliability monitoring and product qualification. Modeling should offer a domain for acceptable number of IST cycles for different types of applications.

An example of IST-based EDOE frame for reliability is shown in Figure 8.1. Central DOE is surrounded by four component designs corresponding to the main factors.

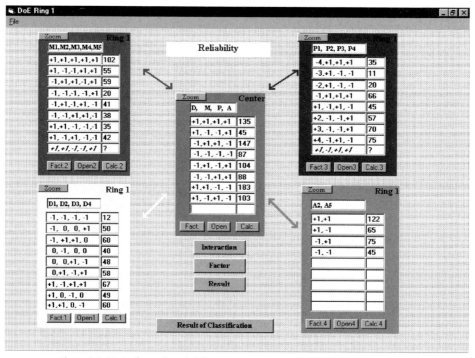

Figure 8.1 EDOE frame for reliability.

Table 8.2 PTH reliability central DOE.

Factor	"−1"	"1"
D-Design	Old IST coupon	IST minicoupon
M-Materials	M- (FR226)	M- (FR404)
P-Processes	"O"	"N"
A-Application	Without assembly	Assembly

O, old; N, new. $D > M > P > A$.

Table 8.3 PTH reliability design-D.

Factor	"−1"	"0"	"1"
D1-Hole size	10–12 mil	12–16 mil	16–18 mil
D2-Board thickness	62–87 mil	50–62 mil	35–50 mil
D3-Pad size/hole size	>2.25	1.5–2.5	<1.5
D4-Cu foil type	>1 oz	1 oz	<1 oz

1 mil = 25 µm; 1 oz corresponds to 35 µm plated copper.

The objective function was the IST lifetime. Table 8.2 summarizes the factors of the central DOE. The associated matrix is W8,4,2a.

Table 8.3 contains the D-design factors.

The DOE matrix for D is L9,4,3b and is included in Figure 8.1.

In this case, based on real Fourier analysis, the factor hierarchy is $D4 > D2 \geq D3 \geq D1$. Table 8.4 contains the M-material factors.

The DOE matrix is W8, 5, 2a and is shown in Figure 8.1.

For these data, $M3 \geq M2 = M4 = M5 \geq M1$.

Table 8.5 contains the processing-P factors.

The DOE matrix is of the type W8, 4, 8 and is included in Figure 8.1.

For these data, $P1 > P2 > P4 > P3$. A new ranking of factors should be performed, which will facilitate the occurrence of corrective actions. These correctives should be done in the order of correlation for factors. Processing the coupon through an assembly simulator may reproduce the application-A. The test matrix is W8, 4, 2a.

Table 8.4 PTH reliability materials-M.

Factor	"−1"	"1"
M1-Resin T_g	120–135 °C	135–155 °C
M2-Cu Elongation	10–17%	17–22%
M3-Z-expansion	4.7–5.5%	4.7–4.4%
M4-Delamination time	5–10 min	10–20 min
M5-T_g CV	0.03–0.04	0.01–0.03

Table 8.5 PTH reliability processing-*P*.

Factor-Cu thick setting	−4	−3	−2	−1	1	2	3	4
P1-Copper thick, mil	0.3	0.5	0.7	0.9	1.2	1.5	1.7	1.9
Factor	"−1"				"1"			
P2- Drilling quality	FHS: 10–24 mil				FHS > 24 mil			
P3-Surf finish type	HASL				ENIG			
P4- Copper thick, CV	High				Low			

CV, coefficient of variation; ENIG, electroless nickel immersion gold; FHS, finite hole size; HASL, hot air solder leveling; 1 mil = 25 µm

The matrix and the results are included in Figure 8.1. For this particular experiment, it shows that the more significant influences on the lifetime have the thermal excursions to which the coupon is exposed during assembly simulation. Detailed information concerning the application would be obtained at the next level of DOE. Two tests have been considered here. They are standard tests in the practice of testing reliability for automotive applications on engine ("−1") and off engine ("1"). Solder shocks test is conducted by dipping the coupons in solder at 288 °C (10 s in solder, followed by 50 s at room temperature). Air-to-air (A/A) tests consist of 1000 cycles with a low temperature of −40 °C and a high temperature of +125 °C (that is, Auto-B) or of 1000 cycles with a low temperature of −40 °C and a high temperature of +145 °C (that is, Auto-C). The dwell time is 25 min, and the transition time is 5 min.

The coupons have been stressed according to tests shown in Table 8.6 and then tested by IST machine. The DOE matrix and the IST lifetimes are presented in Figure 8.1. The result is that the two reliability tests are not very different but the air-to-air test is more useful for reliability purposes. Particular forms of the EDOE frame can be easily developed, since IST is an accelerated test.

The resulting EDOE frame facilitates simulations and predictions. New vectors, that is, new settings of parameters might enter as new rows in DOE matrices and will be classified. Consider that *D* and *A* are maintained as "1" in the center design. A forward step is simulated for *M* and *P* only. Consider the materials-*M* vector $\langle 1, 1, -1, -1, 1 \rangle$ (Figure 8.1). This is classified in the same class as the run 7 in matrix *M*. This allows predicting a lifetime of 35 cycles. Suppose that the processing *P* vector will be $\langle 1, -1, 1, 1 \rangle$ (Figure 8.1). This potential experiment will be classified in the same class as the run 5 in the *P*-design. This allows a prediction of a lifetime close to 45 cycles. If it is to associate one digit with the new experiments, these will be "−1" for *M* and "−1"

Table 8.6 PTH reliability application-*A*.

Factor	"−1"	"1"
A2-S/S	10	5
A5-A/A	Auto-C	Auto-B

for P, since this corresponds conventionally to the low lifetime. This illustrates the transition from real data as the lifetimes to digital data as the DOE elements.

The backward step for the new test, where D and A have been maintained but M and P have been modified, involves joining the vector $\langle 1, -1, -1, 1 \rangle$ to the center design. This is in fact the second run here and allows predicting a life of 45 cycles. Observe that a calculus of similarity should be performed for the backward steps. Similarities, such as defined in Chapter 5, replace probabilities and significantly simplify calculations.

8.2
SKUP Schema for Reliability

The schema presented in Figure 8.2 illustrates elements of the *SKUP* schema linked to reliability of EDOE shown in Figure 8.1. The *SKUP* contains the elements $S = (s^0, s^1)$; $K = (k^0, k^1)$; $U = (u^0, u^1)$; and $P = (p^0, p^1)$. These elements correspond to the two levels, $m = 0$ and 1.

Typical notations are D-design, M-materials, P-processes, and A-application corresponding to the level $m = 0$. D1, D2, D3, and D4 are the factors of D; P1, P2, P3, and P4 are the factors of P that correspond to the level $m = 1$; A2 and A5 are the applications.

The EDOE presentation is limited to detailed study of D and P.

Factors D, M, and so on are the conditions at the level 0. Let $D = k_0^0$, $M = k_1^0$, $P = k_2^0$, and $A = k_3^0$. The system initial state is s_0^0. The evolution at the level $m = 0$ is illustrated by high-thickness border cells.

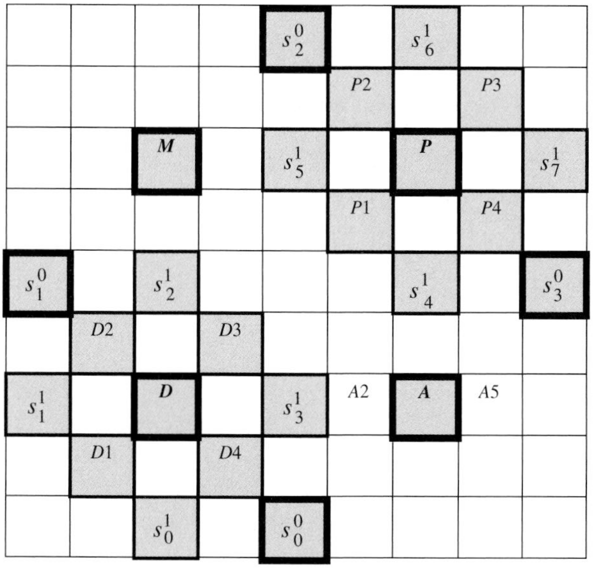

Figure 8.2 *SKUP* schema for reliability.

With possibility $p^0(k_0^0|s_0^0)$, the condition k_0^0 is selected.

This is a digit symbolizing a specific design. Based on this, the operator u^0 defined by $s_1^0 = u^0(k_0^0, s_0^0)$ allows the transition to the new state s_1^0. Then, with possibility $p^0(k_1^0|s_1^0)$, the new condition k_1^0 arises. This condition symbolized by a digit corresponds to the selection of materials. Based on the operator $u^0(k_1^0, s_1^0) = s_2^0$, the system reaches the state s_2^0 and so on.

If experiments show that D and P are significant factors, the analysis may be continued at the level $m = 1$ for the following test conditions: $D1 = k_0^1$, $D2 = k_1^1$, $D3 = k_2^1$, $D4 = k_3^1$, $P1 = k_4^1$, $P2 = k_5^1$, $P3 = k_6^1$, and $P4 = k_7^1$. The trajectory at the level $m = 1$ is illustrated by medium-thickness border cells. The system initial state at the level $m = 1$ is s_0^1. With possibility $p^1(k_0^1|s_0^1)$, the condition k_0^1 arises. This is a digit symbolizing $D1$. Based on this, the operator u^1, defined by $s_1^1 = u^1(k_0^1, s_0^1)$, describes the transition to the new state s_1^1 and so on.

With possibility $p^1(k_1^1|s_1^1)$, the condition k_1^1 arises. This is a digit symbolizing $D2$. Based on this, the operator $u^1(k_1^1, s_1^1) = s_2^1$ describes the transition to the new state s_2^1.

The process may continue as shown in Figure 8.2.

The interpretation of the high-thickness border cells trajectory is as follows: from the state s_0^0 through condition k_0^0 toward the state s_1^0, then through condition k_1^0 toward the state s_2^0, and so on. The circular schema is $(s_0^0, k_0^0) \rightarrow (s_1^0, k_1^0) \rightarrow (s_2^0, k_2^0) \rightarrow (s_3^0, k_3^0) \rightarrow (s_0^0, k_0^0)$.

The self-similar or nested character of EDOE construction is evident. The medium-thickness border cell trajectories reproduce the high-thickness border cell trajectories and may be operated recursively by the same soft.

The loop at the level $m = 0$ is completed by small loops at the level $m = 1$.

8.3
Reliability Management System: Main Elements

Important features of an operational reliability management system (RMS) are

- Knowledge of the functional and reliability requirements of the product
- Understanding of the operating and environmental condition of the product
- Probable failure modes and mechanisms
- Acceleration model for each failure mechanism
- Methodology for designing and conducting the test, such as EDOE
- Capability to analyze physical failures to interpret results and to draw conclusions.

The architecture of the RMS is based on EDOE methodology. RMS appears as a cognitive network that includes the central DOE, the component DOE, the associated software, and the interconnections between several DOE. DOE steps alternate with the measurements and analyze the steps followed by DOE reorganization. The analysis is done by specific software tools that allow to associate continuous values with the discrete, that is, logical settings from DOE matrices and then again discrete values to continuous results by classification and threshold logic. The RMS includes the software allowing the reliability predictions, the test minicoupon design, the software for failure mode identification by classification, and the tests correlation software. For new products, it is necessary to addresses the problem of reliability evaluation and failure mode database grounding from the viewpoint of anomaly detection and the autonomous failure mode generation. Important questions are as follows: How does DOE detect the newness of an unforeseen situation, which was not explicitly taken into account within the design? How does the control system detect an anomaly? The EDOE frame emphasizes a continuous distinction between model-driven expectations and the reality based on accelerated test. EDOE should have a scanning function that includes recognition of the familiar events and a strategy for unprogrammed events. Large differences between expectation and reality indicate that a new reliability assessment possibly including new failure mode and supplementary analysis is required. It is expected that the modified EDOE will correspond more accurately to the recorded data.

The RMS method presented here is based on the following steps:

- A new product receives information from D, M, P, and A as required by prediction software. Calculate free parameters for imposed application or feasible application for imposed parameters.
- Design the minicoupon that will be implemented on the product frame.
- Based on prediction software results, the correlation software offers the number of IST cycles for the specific application. The number of cycles for other accelerated tests such as S/S and L/L results.
- Run the IST tests. Compare the resulting lifetime with the predicted one.
- Evaluate by classification software the failure modes.
- Implement corrective actions if the failures are of classical type.
- Perform detailed failure analysis if the failure pattern is "new."
- Include the new results in the EDOE frame.
- Run simulations of different virtual designs and perform an analysis of simulation results.
- Establish the directions for reliability improvements.

The key elements of the RMS are briefly described next.

8.4
Reliability Prediction Software

This software involves design rule review from reliability point of view. For the software for predictions, D, M, P, and A elements are input data (Iordache, 1998a, 1998b). The result shows the product request for reliability level and relates to application class. Application class refers to reliability tests, assembly simulation tests, and also field conditions. If the application is indicated, the processing parameters such as the Cu thickness result. The main reliability factors are

- Design, *D*: drilled hole size, board thickness, pad-size ratio, and Cu foil type
- Materials, *M*: phase transition temperature (T_g), the coefficient of variation for T_g (T_g CV), *Z*-expansion, and delamination time at 260C (*T260*).
- Processing, *P*: electroplated Cu thickness, electroplated Cu elongation, electroplated Cu elongation CV (coefficient of variation), surface finish type, and parameters.
- Application: conventional tests such as A/A, IST, L/L, and S/S.

Figure 8.3 shows an example of prediction, that is, the specified parameters and the predicted reliability level.

The data may be based on DOE and SPC applied in succession. If the data are missing, they are replaced by default and by the more detrimental values available. The strategy is "quasilinear," since different contributions are summed up linearly. The example presented in Figure 8.3 neglects interactions. This is a limitation since it is risky to predict how the system will work by understanding factors separately and combining them additively. To take into account nonlinearity, different damage addition rules have been analyzed (Iordache, 1998a, 1998b). Reliability prediction may be a globally ineffective method. Experience concerning design, materials, manufacturing, and field operation should be included in a forward step, from the entire systems to elements, to be able to achieve higher reliability of the electronic system.

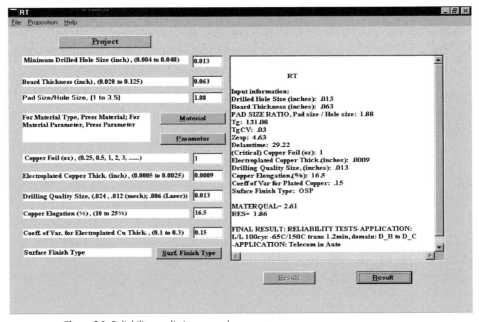

Figure 8.3 Reliability prediction example.

8.5
Minicoupons

The test is done on specially designed minicoupons that are generated automatically (Figures 8.4 and 8.5). Figure 8.4 shows the circuits and the main parameters of the minicoupons.

They are

- Diameters of the holes (*A* predominant, *B* the smallest in part number design)
- Diameters of pads.

Figure 8.6 shows a real minicoupon.

The minicoupons are embedded on the frame of production panels. For the case presented here, the heating temperature is 200 °C, the cooling is 25 °C, and the dwell

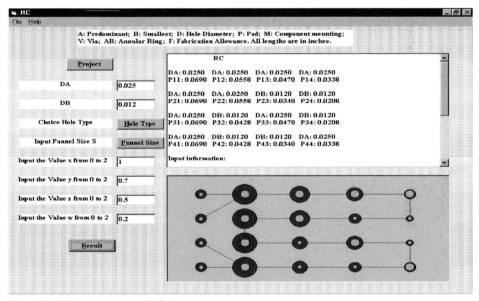

Figure 8.4 Minicoupon automatic design.

Figure 8.5 IST minicoupon for reliability.

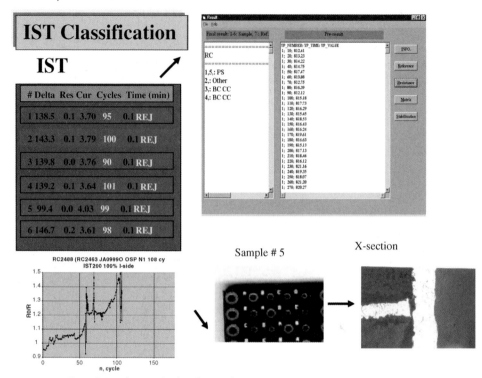

Figure 8.6 Failure mode identification by IST.

time is 3 min. The reject criterion is at 10% of the initial electrical resistance. The number of thermal cycles to failure and the electrical resistance are recorded. The correlation model offers a domain for acceptable number of IST cycles for different types of applications.

8.6
Reliability Analysis

The failure classification software implementation concentrates on identifying the type of failure. Correlation between the electrical resistance variations in minicoupon as measured by IST testing technologies and the failure types in the whole interconnection is necessary. The evaluation of minicoupon reliability is viewed as a process of pattern classification.

An example of IST test results with IST life and resistance recording is shown Figure 8.6. On the basis of IST patterns, the types of failures are identified. They may be postseparations (PSs), barrel cracks (BCs), corner cracks (CCs), or others.

The nonclassified failures impose new microsections.

The lifetime is compared with the predicted ones. Figure 8.7 shows an example.

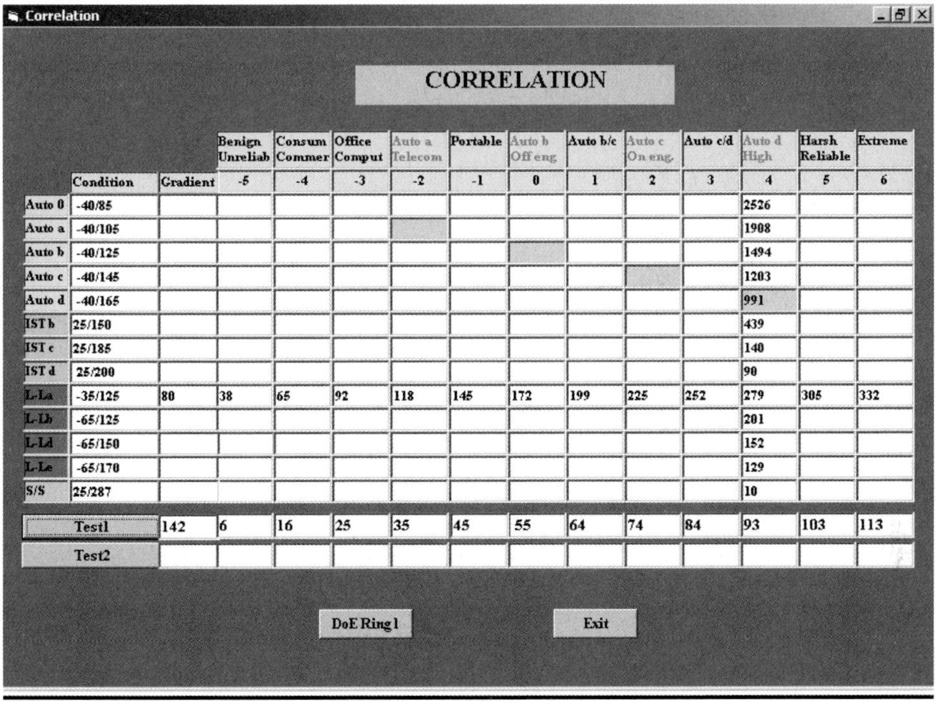

Correlation ▯ _ 🗗 ✕

CORRELATION

			Benign Unreliab	Consum Commer	Office Comput	Auto a Telecom	Portable	Auto b Off eng	Auto b/c	Auto c On eng	Auto c/d	Auto d High	Harsh Reliable	Extreme
	Condition	Gradient	-5	-4	-3	-2	-1	0	1	2	3	4	5	6
Auto 0	-40/85											2526		
Auto a	-40/105											1908		
Auto b	-40/125											1494		
Auto c	-40/145											1203		
Auto d	-40/165											991		
IST b	25/150											439		
IST c	25/185											140		
IST d	25/200											90		
L-La	-35/125	80	38	65	92	118	145	172	199	225	252	279	305	332
L-Lb	-65/125											201		
L-Ld	-65/150											152		
L-Le	-65/170											129		
S/S	25/287											10		
Test1		142	6	16	25	35	45	55	64	74	84	93	103	113
Test2														

DoE Ring1   Exit

Figure 8.7 Correlation.

In the correlation presented in Figure 8.7 the number of cycles corresponding to different test conditions may be compared (Iordache, 1998b; Iordache and Hu, 1999).

8.7
IST Electrical Resistance Analysis

The previous approach is based on the use of similarities between vectors linked to each diagram of electrical resistance versus time, processing of IST minicoupons, as shown in Figures 8.8–8.16. These figures illustrate the main patterns of resistance as a function of the number of cycles.

For a good product, the pattern should be one of the known failure modes such as the barrel crack shown in Figure 8.10 or the corner crack shown in Figure 8.11.

The anomaly such as the inner-layer (I/L) separation, the postseparation, should be recognized. The resistance versus number of cycles is coded as a vector. Each vector corresponds to a specific resistance versus the number of cycles diagram and this in turn corresponds to a tested sample. Matrices of similarity and algorithms and grids of classification have been established.

To distinguish between conventional and nonconventional failures, the algorithm must ensure that the comparison between sets of patterns is carried out in the intended way. There are several distances between database and new patterns to be

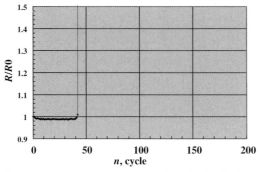

Figure 8.8 Resistance versus the number of cycles for barrel crack.

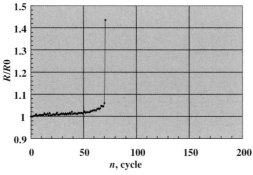

Figure 8.9 Resistance versus the number of cycles for corner crack.

Figure 8.10 Barrel crack.

tested. It is necessary to learn the appropriate distance, the acceptable degrees of similarity, and so on. The classification given in Table 8.7 presents an example. The classification algorithm is as follows: two patterns are assigned to the same class at the grouping degree T, $0 < T < 1$ if their similarity in the stable matrix is larger than T.

Figure 8.11 Corner crack.

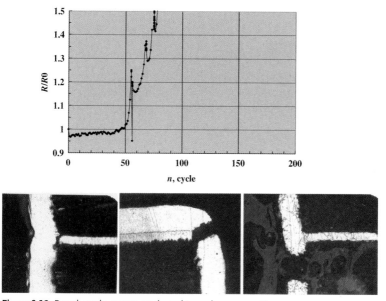

Figure 8.12 Barrel crack, corner crack, and inner-layer separation.

The italicized vectors from {1} to {4}, correspond to the examples retained for classification.

It is a matrix with four rows and eight columns. The vectors {5}, {7}, and {X} represent the samples to be classified. The Ni/Au-plated samples as well as the soldered samples are classified in the same class as the barrel cracks. Both samples should show after microsectioning the barrel cracks. The unknown sample denoted by X pertains to the same class as the vectors {3} and {4}. The cross section outlined I/L separation.

Table 8.7 shows evidence for two types of memory in classification tables. The first four patterns pertain to long-term memory, while the following patterns correspond

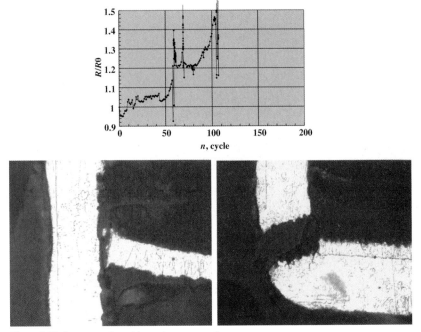

Figure 8.13 Corner crack and inner-layer separation.

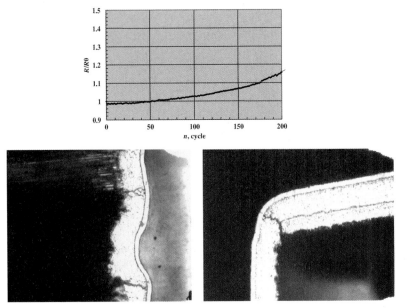

Figure 8.14 Nickel/Gold plating.

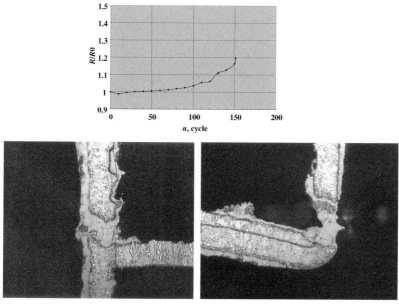

Figure 8.15 Soldered PTH.

to short-term memory. After coding, the pattern vector enters in the short-term memory. Some information is lost by forgetting, but the one which is retained is transferred to the long-term memory, where it is organized in memory structures as classification tables. The knowledge database manages both static and dynamic information. Systems with permanent memory have been developed also. Recogni-

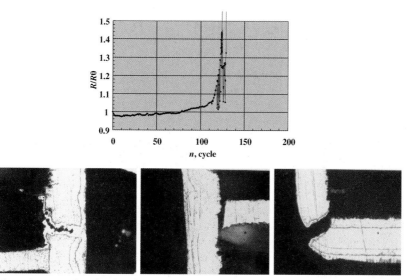

Figure 8.16 Sample to be classified.

Table 8.7 Resistance patterns: classification table.

#	Table	Failure	T = 0.8	T = 0.9	T = 0.95	Figure
{1}	1−1−1−1 1−1−1−1	BC	{1, 5, 7}	{1, 5, 7}	{1}	8.8
{2}	−1−1−1−1 1 1−1−1	CC	{2}	{2}	{2}	8.9
{3}	1 1−1−1−1 1−1−1	BC, CC, PS	{3, 4, X}	{3, 4, X}	{3, 4, X}	8.12
{4}	1 1−1−1 1 1−1−1	CC, PS	{3, 4, X}	{3, 4, X}	{3, 4, X}	8.13
{5}	1−1−1−1−1 1 1 1	Ni/Au	{1, 5, 7}	{1, 5, 7}	{5, 7}	8.14
{7}	1−1−1−1−1 1 1−1	Soldered	{1, 5, 7}	{1, 5, 7}	{5, 7}	8.15
{X}	1 1−1−1−1 1 1−1	BC, CC, PS	{3, 4, X}	{3, 4, X}	{3, 4, X}	8.16

BC, barrel crack; CC, corner crack; PS, postseparation (I/L separation).

tion processes may be considered as a kind of memory; all that the task implies is the ability to advertise what is perceived has been met before. Simply including the pattern produced by the classification algorithm within the database does not solve the problem. The resolution of sensors can be arbitrarily increased, so that they do not only determine whether some resistance dependence is a yes/no pattern but also detect fine differences between a "no" or "yes" class for the product. Perceptive frame should also include the evaluation of reliability, that is, the system action. This is important because a change in pattern does not necessarily indicate problems unless it is coupled with an important change in external behavior. The classification software may be supplemented with a set of corrective actions. For instance, if the I/L separation is detected, the set of corrective actions is described according to the failure severity. It is possible to extend the frame to include other software components as decision-making once a failure mode is detected. In particular, those components on which the observed anomaly detection relies should have priority. The RMS needs to be periodically checked and updated. The method of software monitoring may involve the checking of expected signatures for failures with actual patterns and concentrates on monitoring of anomaly detection. Modification in the control software may result in an anomalous behavior, but it may still continue to maintain the functioning level as required. It is necessary to establish what is important for survival. For some electronic devices, maintaining minimal reliability is the critical requirement.

9
Drilling

9.1
Drilling Quality Framework

The drilling process in the manufacture of PCB is one of the operations where there is an extensive variation in the parameters used. The reason for this is the product diversity and the variation in the definition of what is required (Goulet, 1992). The drilling for PCB industry is a subject of continuous qualitative changes.

PCB produced using conventional fabrication technology has pad sizes and through holes that cannot support the new generation of electronic components. Designs are required to be smaller, more compact, and more reliable.

Microvia is the key technology intended to accommodate high-density interconnections. Microvia has been defined as any via that is 150 μm, that is, 6 mil, or less in size before plating. This definition includes blind/buried via as well as through holes. The quality of microvia effectively depends on design, on materials, on processing steps (such as drilling, imaging, and plating), and on controls and testing. Specific technologies in production or under-development include photovia, laser ablation, plasma etching, and mechanical drilling. The quality of a drilled hole through a PCB is measured by its ability to interface with plating and solderability and form a highly reliable nondegrading electrical and mechanical connection.

The proposed drilling quality framework is shown in Table 9.1. As a first trial, the point of view that the drilling process-related parameters are more significant than PCB-related features was accepted.

D, M, P, and A pertain to center design. Factors indexed by only one number pertain to the primary levels in EDOE, whereas factors indexed by two numbers pertain to the secondary levels in EDOE.

9.2
Test Coupons

A variety of test patterns allowing determining the effects of different quality conditioning levels were designed and tested in PCB industry. Significant was the

Evolvable Designs of Experiments: Applications for Circuits. Octavian Iordache
Copyright © 2009 WILEY-VCH Verlag GmbH & Co. KGaA, Weinheim
ISBN: 978-3-527-32424-8

Table 9.1 PCB drilling framework.

Design-*D*	Materials-*M*	Processing-*P*	Application, control-*A*
*D*1-drill design	*M*1-drill materials	*P*1-drilling machine	*A*1-air–air 1000 h
*D*2-holes	*M*2-entry and *M*2-backup	*P*2-drilling process	*A*2-liquid–liquid
*D*3-board geometry	*M*3-laminate	*P*3-lamination	*A*3-IST
		*P*4-hole condition	*A*4-optical
		*P*5-plating	
*D*1-drill design	*M*1-drills materials	*P*1-drilling machine	*A*1-air–air 1000 h
*D*1.1-drill geometry	*M*1.1-carbide type	*P*1.1-vacuum	*A*1.1-electrical
*D*1.2-drill surf. Fin	*M*1.2-drill condition	*P*1.2-vibration	*A*1.2-microsection
*D*2-holes	*M*2-entry/back	*P*1.3-spindle	
*D*2.1-hole size	*M*2.1-resin type	*P*2-drilling process	*A*4-optical
*D*2.2-hole density	*M*2.2-hardness	*P*2.1-speed	*A*4.1-delamination
*D*2.3-hole location	*M*2.3-flatness	*P*2.2-feed	*A*4.2-nailheading
*D*2.4-hole type	*M*3-laminate	*P*2.3-hits/drill	*A*3.3-smear
*D*3-board geometry	*M*3.1-resin	*P*2.4-loads	*A*4.4-burrs
*D*3.1-thickness	*M*3.2-copper	*P*2.5-pressure foot	*A*4.5-debris
*D*3.2-nr. layers	*M*3.3-glass/resin	*P*2.6-stack height	*A*4.6-roughness
*D*3.3-construction	*M*3.4-oxide	*P*2.7-time btw. hole	*A*4.7-loose fibers
		*P*2.8-retract rate	
		*P*3-lamination	

D > *M* > *P* > *A*; *D*1 > *D*2 > . . . , *M*1 > *M*2 > . . . , *P*1 > *P*2 > . . . ,
*D*1.1 > *D*1.2 > *D*1.3 > . . . > *D*2.1 > *D*2.2 > *D*2.3 > . . . , *M*1.1 > *M*1.2 > *M*1.3 > . . . ,
*T*1.1 > *T*1.2 > *T*1.3 > . . . .

embedding of DOE matrix in the test coupons (Iordache and St. Martin, 1997, 1998).

As in conventional statistical DOE, two tendencies compete in the design of test patterns. First, it is possible to consider one-factor-at-a-time or other unstructured method of experiment design. This cannot capture the essential nonlinear character of a complex drilling process and the interactions of factors. Another method utilized here is to include as much as possible information in a minicoupon, as in fractional designs. Test coupon designs have been based on solutions of the first-order wave equation (WE).

Table 9.2 corresponds to a test pattern developed during a preliminary study of drilling. The test pattern includes the parameters: hole type, drilling speed, and drill age. The notations are as follows: first digit – the hole type; second digit – drill types; and third digit – the drill age. Each parameter has three settings corresponding to low, average, and high.

The central cell from Table 9.2, indexed by 1, 2, 0, contains average hole size and high drilling rate and was done with low age drills.

Table 9.2 Test pattern for drilling.

0 0 0	0 1 2	0 2 1
1 1 1	1 2 0	1 0 2
2 2 2	2 0 1	2 1 0

9.3
Testing Small Plated Through Holes: *SKUP* Schema for Drilling

EDOE method was implemented to select the significant factors of design-D, materials-M, processing-P, and application-A. The first row in the drilling framework corresponds to the elements of the EDOE center. Here, D, M, P, and A are the primary levels. For the secondary levels of DOE, the elements indexed by ($D1$, $D2$, $D3$, ...) or ($M1$, $M2$, $M3$, ...) have been considered. For the tertiary levels, the elements indexed by two indices will be grouped in DOE, for example, ($P21$, $P22$, $P28$, ...). The tests allow direct comparison between mechanically drilled via and microvia.

The test coupon from Figure 9.1 was developed as a tool allowing to establish the reliability impact of the mechanical drilling process and to improve the drilled hole quality. The Figure 9.1 shows the internal circuit as a daisy chain and the external power circuit. CTB denotes the cell test board designated to be electroplated.

Table 9.3 contains the initially selected drilling factors Fi and the meaning of settings, "-1" for supposed lower quality and "1" for supposed higher quality. They have been selected as the more significant factors after running the central DOE and some outer level DOE. The center design involved DMPA. Detailed designs at level $m = 1$ have been performed. They outlined $D1$, $D2$ for design, $M1$, $M2$ for materials, and $P2$ for processing as more significant. At the level $m = 2$, detailed designs outlined $P23$, $P24$, and $D32$ as more significant.

The selected factors pertain to secondary levels (denoted $F1$, $F3$, $F4$, and $F6$) and to tertiary levels (denoted $F2$, $F5$, and $F7$). An interaction type of DOE with seven factors was completed.

Figure 9.2 shows the *SKUP* schema associated with drilling.

The *SKUP* schema from Figure 9.2 implies at the level $m = 1$ the conditions $D1 = k_0^1$, $D2 = k_1^1$, and so on. The states and conditions at the level $m = 1$ are illustrated by medium-thickness border cells.

The system initial state at the level $m = 1$ is s_0^1. With possibility p^1 $(k_0^1|s_0^1)$ the condition k_0^1 arises. This is the digit symbolizing $D1$. On the basis of this condition,

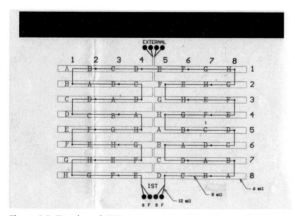

Figure 9.1 Test board CTB.

Table 9.3 Quality factors for drilling, DMP interaction.

Factor	F_i	Noted	"−1"	"1"
Drill material	F1	M1	MF10	Coated DLC
Drill condition	F2	P2.3	Regrind 6 K	New
Drill design	F3	D1	Conventional	Modified
Hole size	F4	D2	6 mil	10 mil
Number of layers	F5	D3.2	High	Low
Entry	F6	M2	EO+	LE Sheet
Chip load	F7	P2.4	0.4	0.2

1 mil = 25 µm.

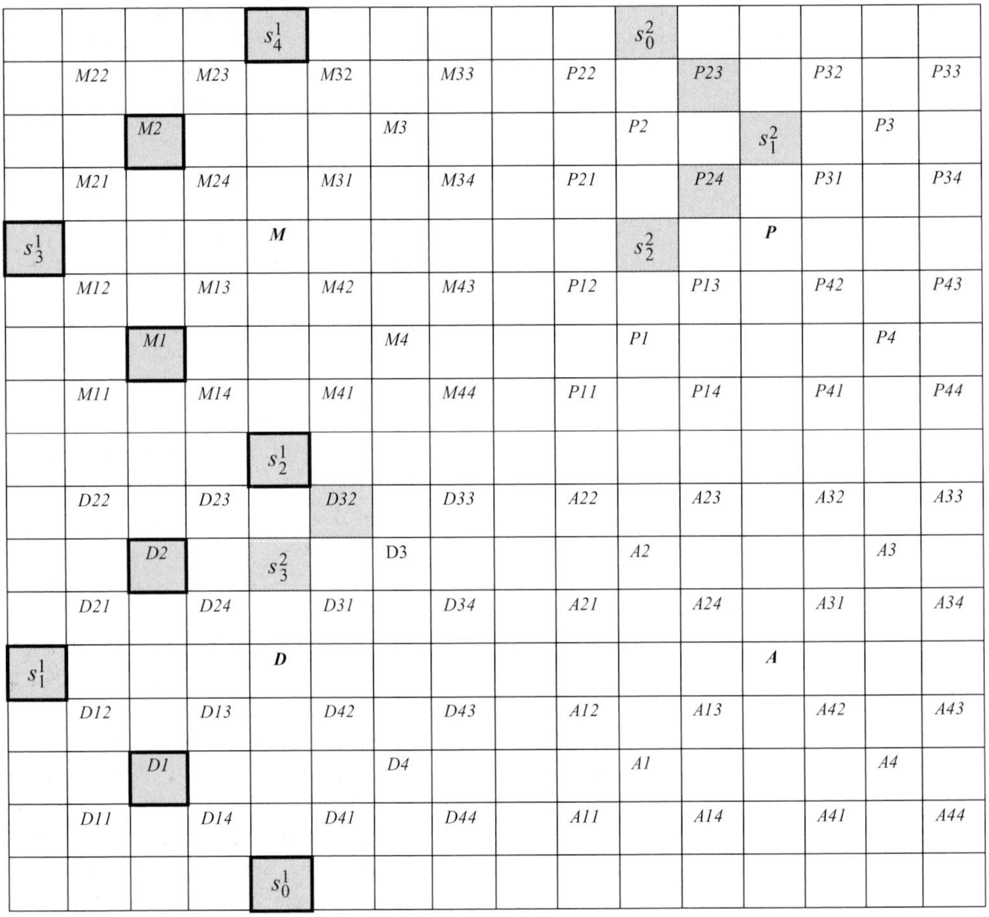

			s^1_4							s^2_0					
	M22		M23	M32		M33		P22		P23		P32		P33
		M2			M3				P2		s^2_1		P3		
	M21		M24	M31		M34		P21		P24		P31		P34
s^1_3			*M*						s^2_2		*P*				
	M12		M13	M42		M43		P12		P13		P42		P43
		M1			M4				P1				P4		
	M11		M14	M41		M44		P11		P14		P41		P44
			s^1_2											
	D22		D23	D32		D33		A22		A23		A32		A33
		D2		s^2_3	D3				A2				A3		
	D21		D24	D31		D34		A21		A24		A31		A34
s^1_1			*D*								*A*				
	D12		D13	D42		D43		A12		A13		A42		A43
		D1			D4				A1				A4		
	D11		D14	D41		D44		A11		A14		A41		A44
			s^1_0											

Figure 9.2 *SKUP* schema for drilling.

Table 9.4 Matrix DMP of DOE.

#	F1	F2	F3	F4	F5	F6	F7
A	1	1	1	1	1	1	1
B	1	1	1	−1	−1	−1	−1
C	1	−1	−1	1	1	−1	−1
D	1	−1	−1	−1	−1	1	1
E	−1	1	−1	1	−1	1	−1
F	−1	1	−1	−1	1	−1	1
G	−1	−1	1	1	−1	−1	1
H	−1	−1	1	−1	1	1	−1

the operator $u^1, u^1(k_0^1, s_0^1) = s_1^1$ describes the transition to the new state s_1^1 and so on. With possibility $p^1 (k_1^1 | s_1^1)$, the condition k_1^1 arises. This is a digit symbolizing D2. On the basis of this, the operator $u^1(k_1^1, s_1^1) = s_2^1$ describes the transition to the new state s_2^1 and so on as shown in Figure 9.2.

At the level $m = 2$ the conditions are $k_0^2 = P23$, $k_1^2 = P24$, and $k_2^2 = D32$. The states and the conditions at the level $m = 2$ are illustrated by standard-thickness border cells. An example of operator is $u^2(k_1^2, s_1^2) = s_2^2$. As observed, relatively few of the positions are activated, but the entire frame may be useful for supplementary experiments. The *SKUP* schema retains and integrates only the conditions and the states that cooperate with its functioning. The situation may be compared with the one encountered in the study of Piaget's action schema or Uexküll's functional cycles.

Figure 9.2 shows the complete symbolic description of the systems for two activated levels.

It is considered as a first trial that the significance order for factors is $F1 > F2 > \cdots > F7$. Specific problems have been outlined in designing the coupon as to how to select the set of significant patterns for via placements on the coupon. The matrix of the proposed screening DOE shown in Table 9.4 is in fact a WH matrix. A, B, ..., H denote the eight possible experiments. For instance, B denotes the experiment $B = [1\,1\,1\,-1\,-1\,-1\,-1]$ meaning coated drill, new drills, modified point angle, 6 mil (150 μm) hole size, and so on. Any drilled hole in the CTB design corresponds to an experiment from Tables 9.4 and 9.5.

Table 9.5 Coupon CTB.

A	B	C	D	E	F	G	H
B	A	D	C	F	E	H	G
C	D	A	B	G	H	E	F
D	C	B	A	H	G	F	E
E	F	G	H	A	B	C	D
F	E	H	G	B	A	D	C
G	H	E	F	C	D	A	B
H	G	F	E	D	C	B	A

The placement of holes, that is, of experiments, on CTB follows an *L* type of matrix. As WH matrices, the *L* matrices are particular solutions of the wave equation. Supplementing the seven selected factors in *L* matrix, two secondary factors need to be considered: the plating thickness, due to the fact that the CTB is a cell test board plated in the Gornall cell, and the hole location on the coupon. The holes are of central type or at the coupon periphery. The coupon CTB includes two electrical heating circuits, one internal I, relating all the 64 holes, and one external E. It is an embedded fractional design selected according to informational criteria. The matrix from Table 9.4 has an associated informational entropy $H(M) = 34.07$ calculated as in Chapter 5. This is the matrix of an ideal experiment. Due to the practical restrictions, some factors were lost during coupon fabrication. The factor *F6* was performed either as "−1" only or as "1" only. In that case, the informational entropy is $H(M') = 33.72$. A more evident loss of information appeared on some test panels where due to accidental drill problems the holes denoted by D and F were missing. In that case, the truncated matrix of design has the informational entropy $H(M'') = 18.30$. Note that the EDOE method proposed here may perform well on incomplete data, and this is an advantage in practical situations. The informational entropy is the control tool for DOE evolution. The coupons were processed by standard plating process to define the circuitry. The only off-line operation was the drilling.

9.4
Reliability Tests

The test coupons have been submitted to different thermal fatigue tests including, air-to-air 1000 h test (A/A), liquid-to-liquid test (L/L), and IST test. Test conditions are summarized in Table 9.6. The reliability test results depend essentially on thermal profile, on temperature and thermal ramps, on coupons, and on data analysis. The required lifetime is calculated on the basis of a reliability prediction model.

Resistance measurements have been performed during hot periods (IST, L/L). Resulting data are presented in Figures 9.3 and 9.4. Figure 9.3 shows electrical resistance as function of the number of cycles for IST test for different holes. Figure 9.4 shows the electrical resistance as function of the number of cycles for L/L test for different holes. Measurements are performed after any 100 or 50 cycles.

Table 9.6 Conditions for reliability tests.

Test	Cold T_c (°C)	Hot T_h (°C)	Dwell T_c (min)	Dwell T_h (min)	Transit time (min)	Required life (cycle/mil)
IST	+25	+185	1.5	1.5	NA	300
L/L	−25	+105	15	15	0.5	100
A/A	−40	+125	25	25	5	1000

NA: not available; 1 mil = 25 μm.

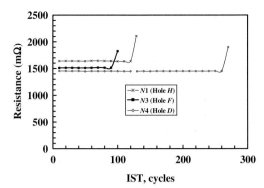

Figure 9.3 Resistance versus number of IST cycles.

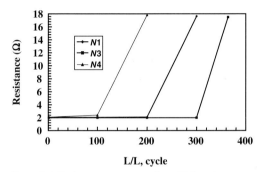

Figure 9.4 Resistance versus number of L/L cycles.

For IST test, the accelerated effects of I-side processing explain the rapid increase of the resistance. For this test panel, the resistance pattern does not allow making inferences concerning the type of failure (type of cracks, I/L-separation).

New theoretical tools have been developed to gain more information from testing. Among these, the lifetime maps allow significant innovations.

9.5
Lifetime Maps

The implemented testing procedure allows to obtain failure maps and lifetime maps. Such maps show the lifetime when hole size, plating thickness, and relative positions are different. The reliability test continues after the repairing of the PTH failure. Repairing is done by filling with silver paste the failed via. The maps highlight the critical areas and are useful like guides for future designs. An example is presented in the Table 9.7.

The uncompleted cells correspond to holes having the lifetime superior to the maximum record (635 cycles in Table 9.7). The hole *F* on the coupon shown in Table 9.7 close to the pass/fail criteria 300 cycles/mil. Table 9.8 contains a recording of the number of IST tests when the corresponding hole failed.

Table 9.7 IST lifetimes.

A	B	C	D	E	F	G	H
A	B	C	D	E	F	G	H
B	A	D	C	F	E	H	G
		222					
C	D	A	B	G	H	E	F
	265		464		368		
D	C	B	A	H	G	F	E
		129				298	
E	F	G	H	A	B	C	D
		635					
F	E	H	G	B	A	D	C
		155					
G	H	E	F	C	D	A	B
H	G	F	E	D	C	B	A

The total number of performed IST tests was 12. The test shows large number of failures for the holes indexed by *D* and *H*. Then *B* and *F* type of hole follows. *B, D, F, H* have been 6 mil (150 µm) holes. Table 9.9 contains an example of lifetime map for L/L test. The holes were tested for time step of 100 L/L cycles. The holes B and D show the worst performances for several L/L tests. The hole *F* meets and exceeds the L/L test requirements.

Table 9.10 contains an example of lifetime map for A/A test. The hole *B* shows low performances.

The best performances have been that for the hole *E*, corresponding to FHS of 10 mil (250 µm). It is the only hole that passes more than 1000 h test. The holes *A, C,* and *G* have been rejected, but usually this happens at 1000 cycle examination.

Table 9.8 Number of IST tests with failed PTH.

A	B	C	D	E	F	G	H
A	B	C	D	E	F	G	H
	1		2				1
B	A	D	C	F	E	H	G
		8		2		7	
C	D	A	B	G	H	E	F
	7		4		7		2
D	C	B	A	H	G	F	E
		5		2		2	1
E	F	G	H	A	B	C	D
					3		1
F	E	H	G	B	A	D	C
		5				2	
G	H	E	F	C	D	A	B
			2		1		
H	G	F	E	D	C	B	A
						1	

Table 9.9 L/L lifetimes.

A	B	C	D	E	F	G	H
B 300	A	D	C	F	E	H	G
C	D 300	A	B	G	H	E	F
D 200	C	B 200	A	H	G	F	E
E	F	G	H	A	B	C	D
F	E	H	G	B 200	A	D	C
G	H	E	F	C	D 200	A	B
H	G	F	E	D	C	B	A

9.6
Drilling Quality Evaluations

For all tests, it is possible to find windows of acceptable parameters, holes, and copper thickness, meeting the test requirements, as indicated in Table 9.10. The method gives more information than a simple pass/fail result. It reports sufficient data to be used for statistical process control.

The quality as resulting from test may be defined in different ways, for example:

- $Q_1 = 1 - (\text{number of failed holes/total number of holes})$;
- $Q_2 = 1 - (\text{number of failed tests/total number of tests})$;
- $Q_3 = \text{lifetime/copper thickness}$.

Table 9.10 A/A lifetimes.

A	B	C	D	E	F	G	H
A	B 230	C	D 160	E	F 460	G	H 1000
B 320	A	D 100	C	F 1000	E	H 460	G 1000
C 1000	D 320	A	B 460	G 1000	H	E	F 1000
D 320	C	B 160	A	H 1000	G 1000	F 460	E
E	F 1000	G	H 460	A	B 320	C	D
F 1000	E	H 1000	G 1000	B 100	A 1000	D 560	C
G	H 560	E	F 460	C	D 230	A	B 320
H 1000	G 1000	F 560	E	D 560	C	B 320	A

The index Q_3 is the required quality criterion. Using the quality index Q_1 and calculating the factor effect, the significance order for factors is $F4 > F2 > F7 > F3 > \cdots$, that is, hole size > drill condition > chip load > drill design > $\cdots$.

The trend shown by the factor $F3$ contradicts the initial expectations. The hole size effect is outlined by all the tests and quality indices as the statistically significant factor. For the majority of tests, the 6-mil (150 μm) hole starts to fail only after the failing of the 10 mil (250 μm) holes. This means that the transition from 6 to 10 mil is critical for the reliability of the technologies involved. The 6 mil (150 μm) is a significant threshold. This reinforces the definition of the microvia domain. However, as the factor analysis holds for the hole F, there exist mechanical drilling conditions that push a 6 mil (150 μm) hole in the quality range for the 10 mil (250 μm) hole size.

Thermal heating temperature plays as a significant role. The failure is frequently in the central area. This is correlated to the failure mode, the barrel crack. The order of quality as shown by the indices Q_1–Q_3 depends on test conditions. Cross-sectioning of failed via was performed to identify the cause of failure. An example of microvia of type B is shown in Figure 9.5. The quality of mechanical drilling for 6 mil (150 μm) FHS is acceptable. Standard electroless/electroplating processing can perform for through hole microvia. The 1.5 oz copper foil may be accidentally overetched. The material shrinkage was excessive for some panels. One ounce corresponds to a 35 μm copper foil.

It can be concluded from this experiment that $F = [-1\ 1\ -1\ -1\ 1\ 1\ -1\ 1]$ gives the best results for 6 mil (150 μm). This hole is processed in the following conditions: drill material MF10, new drills, conventional design, low copper content, that is, no internal layers, and chip load 0.2. The result is a starting point for the new test. This means that the mechanical drilling for the next step in experimental design will be performed in the condition indicated by F. If this is difficult for production purposes, another hole type close to F is selected by applying the classification procedure. Similarities calculated in Chapter 5 may be of help. Close to hole F, as reliability performances, is the hole E. Among the 10 mil (250 μm) holes, E seems to perform better.

The strategy of the experiments is based on EDOE methodology. The first step in DOE, the forward one, allowing the selection of parameters for the DOE matrix, is

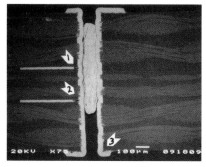

Figure 9.5 Microsection hole type *B*, in CTB.

followed by real data analysis. This step is followed by a backward step in which the less acceptable conditions were discarded, and the search is concentrated on promising directions only. A second forward step in DOE starting from acceptable holes, as *F*, is described in the following sections. After this DOE, the real data analysis is required again. This ensures the EDOE closure and the potential evolvability of the system studied.

9.7
Testing the Design, D

This test permits some direct comparisons to be drawn between laser-defined and mechanically drilled technologies, between one-depth and multidepth microvias. The test minicoupon (TMC) targeted to confirm the feasibility of the technology for drilling and plating studies is developed (Figure 9.6). Figure 9.6 shows the electrical circuit and the TMC. This generic minicoupon is compatible with the major classes of high-density PCB technologies and reliability tests. The objective of the new experiment was to obtain reliability information for blind via technology for production use, based on IST testing technology. Two circuits, an internal one I and an external one E, have been included in this minicoupon. The internal circuit I is a daisy chain heated directly. The E circuit heats externally the tested I circuit.

Studied factors are as follows:

- Type of via, blind via versus through hole.
- Aspect ratio: height *H*/diameter *D*. For one side of the minicoupon $H/D > 3$, multidepth, and for the other side, $H/D = 1$, one depth.

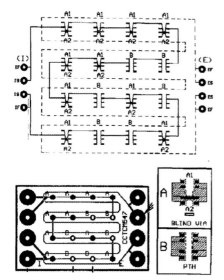

Figure 9.6 Minicoupon, TMC.

Table 9.11 Factors for TMC.

Factor	Fi	"−1"	"1"
Via type	F1	Through hole (TH)	Blind via (BV)
H/D	F2	>1	1

Table 9.12 WH matrices for TMC.

Side 1				Side 2			
A1	A1	A1	A1	A2	A2	A2	A2
A1	A1	B	B	A2	A2	B	B
A1	B	A1	B	A2	B	A2	B
A1	B	B	A1	A2	B	B	A2

- Drilling type, mechanical versus laser, for the same hole.
- Two-step blind via technology versus the one-step blind via makeup.

Supplementing the four selected factors, secondary factors need to be included as follows:

- *Temperature:* This factor was not imposed by design but is a result, as the IST technology heats more of the central area of the minicoupon.
- *Copper thickness:* This depends on the minicoupon position.

Table 9.11 contains the preselected quality factors.

The blind via chains are made up of a total of 16 vias per chain with minimal via diameters, 6 mil (150 μm) by laser. The key features evaluated by the blind via chains are via size and the interconnecting integrity. The entropy calculations for different pattern placement showed that Walsh–Hadamard (WH) type of matrices offer the maximum significance. If four patterns should be tested and there are only two variations for the test pattern, denoted by "A" for blind via and by "B" for through hole via, the WH matrix from Table 9.11 is used.

According to the notations, the vectors from Table 9.12 are $A1 = (1,−1)$, $A2 = (1, 1)$, and $B = (−1,−1)$. In TMC design, A1 or A2 corresponds to blind via and B to through hole via. A1 (A2) was selected as the critical element to be tested more since the design includes 10 holes of type A and only 6 holes of type B.

9.8
Testing for Processing, P

The two-step drilling process, mechanical control depth and laser, was selected for blind via technology. The depth of the drilling at 10 mil (250 μm) hole size was set to

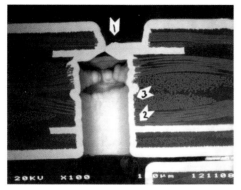

Figure 9.7 Microsection TMC.

penetrate layers 1 and 2. The laser, for 6 mil (150 µm) hole size, finishes the drilling of the layer 3. Graphite direct plate method is used for making via conductive. The direct plate process consists of three major steps: desmear, conductive plate application, black hole in this case, and plating.

The plating was performed with a variable density current. The initial high CD, 8 min at 33 ASF (amperes per square foot, $1\,ASF = 0.108\,A/dm^2$) and 8 min at 30 ASF, induces nucleation of small crystals. The last period of 2 h was at low CD, 11 ASF, for throwing power reason. Electroless nickel/immersion gold surface finish was applied after copper plating in standard conditions. Plating techniques did not seems to be critical for $H/D = 1$ but may become a problem with a greater than 2 aspect ratio. Figure 9.7 shows a microsection representing the minicoupon processing and its quality features. It can be concluded that

- Plating shows nonuniformity on minicoupons close to the panel edge
- Mechanical drilling is acceptable.
- Laser drilling performs better for $H/D = 1$. For $H/D = 3$, the laser induced a specific roughness
- Laser defined via failed for some holes due to insufficient cleaning or residue removal at the bottom of via.

9.9
Reliability Evaluations

The IST test parameters were heating 185 °C, or 200 °C, cooling at 25 °C. The I-side was directly heated. The results have been summarized in the Table 9.13.

Plugging with silver some of vias or only one of the blind vias, the individual reliability may be recorded. SF denotes surface finish (Ni/Au) and PV denotes repaired plugged via. The results show that SF or PV factors are insignificant. Lower lifetimes are due to thickness accidents rather than other parameters. Thin or missing plating is the factor limiting the reliability as shown in Table 9.13. Some of the testing results for TMC are summarized as follows.

Table 9.13 IST reliability results: I-side.

Test	Heat T_h (°C)	Surface finish	Plugged via	IST life cycles
1	185	No SF	No PV	198
2	200	No SF	No PV	87
3	200	No SF	PV	44
4	200	Ni/Au	No PV	90
5	200	Ni/Au	PV	117
6	225	No SF	No PV	73
7	225	Ni/Au	No PV	27

- Initial resistance measurements were collected to analyze the product variability. These are due to process variability, plating, and drilling and represent an evaluation of the technology robustness.

- Failures detected during reliability testing cannot be attributed only to the microvia technology.

- Through hole shows lower reliability than blind via. The failures are preferentially in PTH. This result confirms previous tests showing that microvia will survive thermal cycles and thermal stresses comparably to or better than the through holes. Recall that this well-established concern is not valid if the microvias have nonconformity.

- Thermal heating temperature plays as a significant role. The failure always is in the central area.

- Repairing by plugging via (PV) does not allow to evaluate consistently the reliability of blind via. Repairing was performed by using different types of silver paste and probably was in some case marginal. The repaired blind via lifetime has a large standard deviation. This follows the variability in microvia processing and repairing.

- Ni/Au surface finish is acceptable for any aspect ratio.

- Multidepth blind via technology is less robust especially due to laser-related problems.

- Aspect ratio H/D plays. Higher ratios are less reliable.

The presented evaluation identifies windows of reliable microvia technologies. The mechanically or laser-drilled microvia outlines objective reliability problems. The 6 mil (150 µm) gives acceptable reliability only for specific selection of mechanically drilling parameters. This acceptability refers to theoretically predicted pass/fail criteria. It is a concern that blind microvias are more reliable than through hole microvias. This is valid only for aspect ratio less than 1, if the new problems arising from microvia technology implementation have been solved. In fact, more than the design factors, the processing factors play a limiting role for reliability. The standard PCB manufacturing is not always automatically compatible with microvia technology. Resulting from the reliability tests, the significant critical problems are related to hole conditioning, metallization, throwing power, overetching, and registration.

10
Surface Finish Solderability

10.1
Finish Solderability Frame

Solderability is the ability of a surface to be wetted by molten solder (Wassink, 1989). Good solderability for the PCB as well as for the components has become an important element in achieving the quality required in competitive markets. Microelectronics requires the production of reliable assemblies in an ultralow volume environment. As the new assembly technologies such as ball grid array, flip-chip, and chip on board have progressed, there has been a demand to obtain new solderable surface finishes as an alternative to the conventional hot-air level soldering (HASL). Complex PCBs demand to increase the functionality of the final surface finish (Hwang, 1996). The challenge for printed circuit industry is to correlate the surface finish technology to specific application.

Table 10.1 includes the main factors for the surface finish quality (see Iordache *et. al.*, 1998a, 1998b).

The surface finish selection system is based on EDOE methodology. The central DOE contains PCB design factors, materials factors referring to finish and its application for solderability, processing factors referring to finish, and testing factors referring to solderability after reliability tests. DOE steps alternate with measurements and analyze steps followed by DOE reorganization.

Different tests may be performed to evaluate SF solderability.

Examples of typical tests are

- MUST, wetting balance test
- SERA, sequential electrochemical reduction analysis
- Dip & Look, standard solderability test.

The problem is to quantify the results of all these tests. An example of global criterion that summarizes the partial test significance is introduced next. It associates a real value with a digitalized vector and allows subsequent modification of the DOE matrix.

Denote the test resulting vector for a sample by $S = [i_1, \ldots, i_k \ldots]$. Here, i_k is the digit "1" or "0" corresponding respectively to the result pass or fail of the kth test. To any global test result S, a valuation $V(S)$ defined by $V(S) = \Sigma_k i_k (0.5)^k$ is

Evolvable Designs of Experiments: Applications for Circuits. Octavian Iordache
Copyright © 2009 WILEY-VCH Verlag GmbH & Co. KGaA, Weinheim
ISBN: 978-3-527-32424-8

Table 10.1 Surface finish quality framework.

Design, D	Materials, M	Process, P	Application, A
D1-Pad size	M1-SF type	P1-SF thickness	A1-Thermal cycling
D2-Hole size	M2-SM type	P2-SM application	A2-IST
D3- Heat transfer	M3-Flux type	P3-Contamination	
	M4-Solder type		

SF, surface finish; SM, solder mask.

associated. According to the valuation formula, the solderability for any sample depends on the significance associated in the hierarchical testing sequence with the partial tests. For this case study, the first test has a weight of 0.5, the next 0.25, the following 0.125, and so on, the proposed hierarchy for tests being MUST > SERA Dip and Look > . . ..

The valuation $V(S)$ is in fact a similarity as defined in Chapter 5. $V(S)$ gives similarities relative to a reference vector containing only "1" as coordinates.

The experiment is based on EDOE methodology. DOE steps alternates with measurements and analyze the steps followed by DOE reorganization.

The design-*D* matrix is W4,3,2a and the results are presented in Figure 10.1 (Table 10.2). Figure 10.1 shows the EDOE frame linked to surface finish study. The test minicoupon is shown in Figure 10.2. This minicoupon allows performing all the solderability tests after an imposed number of IST cycles. It evaluates solderability as a function of the testing time.

Preliminary tests assured that materials-*M* and processes *P* factors are significant variables. Table 10.3 contains the notations for *M* settings. ENIG denotes electroless nickel, immersion gold finish. *S/M* type depends on the supplier ("D", "C", or "T").

A DOE matrix of type L9,2,3 with three settings is considered. The main factor is *M2*, that is, the *S/M* type. The processing-*P* factors are included in Table 10.4.

The processing-*P* matrix is of the type W4,2,2a, and the results are presented in Figure 10.1. From this DOE, it results that the factor *P1*, the coating thickness, is the more significant for solderability. An interaction DI experiment for MPA was performed at this stage on the first level of EDOE. The electroless Ni/Au and a compatible mask indexed by "*T*" have been selected to perform an application test. Nine values of the thermal cycling parameters corresponding to the application test have been considered. The time step for the number of cycles is 25 cycles. In this case, −4 is linked to 0 cycles, −3 to 25 cycles, and so on till +4 that is linked to 200 cycles. The corresponding matrices are of the type L*n,m,s*.

Table 10.2 Surface finish: design-*D*.

Factor	"−1"	"1"
D1-Pad size	Small	Large
D2-Hole size	Small	Large
D3-Heat transfer	With	Without

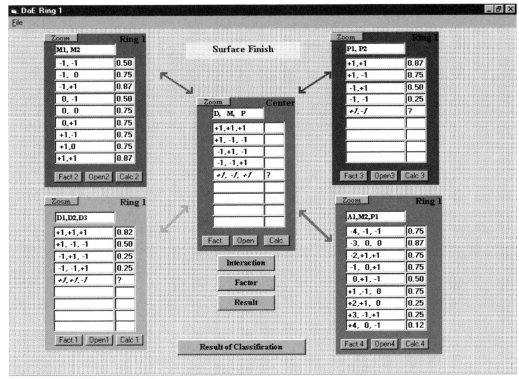

Figure 10.1 EDOE frame for surface finish.

The *M*, *P*, and *A* factors are lumped together in the interaction type of experiment from Table 10.5. This test shows that the application factor, that is, thermal cycling plays significantly. The set of resulting DOE matrices is useful in the implementation of new surface finish technology. The EDOE method allows simulations for new surface finish.

Suppose that the new processing vector will be $\langle 1, -1 \rangle$ (see Figure 10.1). This designation is followed by a forward step. The performed experiment will be classified as the run 2 in processing-*P* matrix. This allows predicting solderability valuation of 0.75. Consider also a new designation step in which the design-*D* vector

Figure 10.2 IST minicoupon for solderability.

Table 10.3 Surface finish: materials-*M*.

Factor	"−1"	"0"	"1"
M1-Surface finish	ENIG(P)	Electroplate Ni/Au	Ag
M2-Solder masks	"D"	"C"	"T"

is $\langle 1, 1, -1 \rangle$. This is classified as the first run in matrix *D*. If high solderability means valuation higher than 0.75, these two vectors that are new experiments will provide digits "1" in the center matrix. These replacements summarize the information and represent backward steps. They translate a real valuation of the solderability into a digit only. Observe that a calculus of the valuation, that is, of similarity allows a shifting from forward to backward steps. Coupled with materials *M* corresponding to lower than 0.75 solderability (that is, "−1" digit in the central matrix), the new vector in the central matrix will be $\langle 1, -1, +1 \rangle$. This is similar to the second run $\langle 1, -1, -1 \rangle$ in the initial center design and it is predicted that this will show performances similar to that run.

10.2
SKUP Schema for Surface Finish

The Figure 10.3 illustrates the *SKUP* schema linked to EDOE frame shown in Figure 10.1.

The *SKUP* contains the elements $S = (s^0, s^1)$; $K = (k^0, k^1)$; $U = (u^0, u^1)$; $P = (p^0, p^1)$. *D*-design, *M*-materials, *P*-processes, and *A*-application correspond to the level $m = 0$; *D1*, *D2*, *D3*, and *D4* are the factors of *D* corresponding to the level $m = 1$. *D*, *M*, *P*, and *A* are the conditions at the level $m = 0$. Let $D = k_0^0$, $M = k_1^0$, $P = k_2^0$, and $A = k_3^0$.

Table 10.4 Surface finish: processing-*P*.

Factor	"−1"	"1"
P1-Coating thickness	High	Low
P2-SM application	After	Prior

Table 10.5 Factors for materials-*M*, processing-*P*, and application-*A*.

Factor	−4	−3	−2	−1	0	1	2	3	4
A1-thermal cycles IST	0	25	50	75	100	125	150	175	200
Factor	"−1"			"0"			"1"		
M2-SM type	"D"			"C"			"T"		
P1-Coating thickness	High			Average			Low		

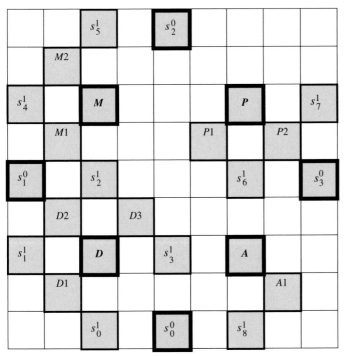

Figure 10.3 *SKUP* schema for surface finish.

The system initial state is s_0^0. The states and the conditions at the level $m=0$ are represented by high-thickness border cells. With possibility $p^0(k_0^0|s_0^0)$, the condition k_0^0 is selected.

This is a digit symbolizing a specific design that is the matrix corresponding to *D*-DOE. Based on this, the operator $u^0(k_0^0, s_0^0) = s_1^0$ allows the transition to the new state s_1^0. Then, with possibility $p^0(k_1^0|s_1^0)$, the new condition, k_1^0 arises. This condition symbolized by a digit corresponds to the selection of materials. In the new condition, the operator $u^0(k_1^0, s_1^0) = s_2^0$ allows the system to reach the state s_2^0. With possibility $p^0(k_2^0|s_2^0)$, the processes k_2^0 are selected and finally, the product $u^0(k_2^0, s_2^0) = s_3^0$ results. It will be tested at the level $m=0$ in the condition $A = k_3^0$. After the test, the state is s_4^0.

The analysis may be continued at the level $m=1$ for the test conditions $D1 = k_0^1$, $D2 = k_1^1$, $D3 = k_2^1$, $M1 = k_3^1$, $M2 = k_4^1$, $P1 = k_5^1$, $P2 = k_6^1$, and $A1 = k_7^1$. The system initial state at the level $m=1$ is s_0^1. The states and the conditions at the level $m=1$ are represented by medium-thickness border cells. With possibility $p^1(k_0^1|s_0^1)$, the condition k_0^1 arises. This is a digit symbolizing $D1$. Based on this, the operator $u^1(k_0^1, s_0^1) = s_1^1$ describes the transition to the new state s_1^1 and so on.

With possibility $p^1(k_1^1|s_1^1)$, the condition k_1^1 arises. This is a digit symbolizing $D2$. Based on this, the operator $u^1(k_1^1, s_1^1) = s_2^1$ describes the transition to the new state.

11
Direct Plate

11.1
Direct Plate Reliability

Direct plates (DPs) are interfaces that make the PCB through holes conductive (Iordache and St. Martin, 1996). The main DP classes are

- Carbon/graphite system
- Palladium system
- Nonformaldehyde-based electroless copper
- Conductive polymer.

The replacing conventional electroless copper process in PCB industry has been evaluated, using information generated by accelerated fatigue test IST and the EDOE methodology for data analysis. The DP technologies were applied in 25 sites, that is, in 25 plants. Evaluation includes electronic microscopy studies of the interfaces between DP, resin-glass, and electroplated copper. Different thermal fatigue processes exhibit different electrical resistance versus lifetime patterns. The classification method used here is based on the similarities between resistances versus lifetime diagrams. Algorithms using similarity matrices based on cellular grids of classification are developed. The general frame is that of EDOE with emphasis on classification methodologies. DOE steps alternate with measurements and analysis steps followed by design reorganization. The analysis is done by specific software tools associating continuous values to the discrete, that is, logical settings from DOE matrices and then again discrete values to continuous results by classification and threshold logic. The EDOE includes the software, the test minicoupon design, and the software for failure mode identification by classification. The results of the study suggest that for any type of DP, at least one site-offering product of high quality is found. By electroless copper deposition, a nonconductive surface, fiberglass, and epoxy can be coated with a conductive layer of copper allowing subsequent electroplating. Environmental issues were an early motivation for the development of alternative plating methods. A second motivation for DP is the increasing demand for low expensive PCB. PCB manufacturer is looking for faster methods of production, while making the process easier,

Evolvable Designs of Experiments: Applications for Circuits. Octavian Iordache
Copyright © 2009 WILEY-VCH Verlag GmbH & Co. KGaA, Weinheim
ISBN: 978-3-527-32424-8

cheaper, and more reliable. There are more classes in which most DP processes can be divided: carbon systems, palladium systems, conductive polymer systems, and nonformaldehyde electroless systems. Different sites have been evaluated using an accelerated fatigue test IST. The heating temperature was 150 °C, the cooling temperature was 25 °C, the dwell time was 3 min, and the rejection criterion was at 10% of the electrical resistance. The numbers of thermal cycles to failure and the electrical resistance have been recorded.

11.2
Microsectioning Results

Microsectioning and visual examination to outline where and how the interconnection Cu foil-DP or electroless Cu-electroplated Cu failed, following the fatigue test. Microsections of minicoupons, processed by IST, was examined using a scanning electron microscope (SEM) equipped with energy dispersive X-ray analyzer (EDS).

Some results and observations are presented in Table 11.1.

The Table 11.1 contains data such as the panel or the sample number as indicated by an independent laboratory, the test site and the DP type, the electroplated copper thickness, and the degree (percentage) of I/L-separation as resulting from interconnection evaluation by SEM. One proposed the degrees of I/L separations: "1", no I/L separation; "0", developed I/L separation or incomplete I/L separation. I/L separations are of the type C (between electroless or DP and electroplated plated Cu), type F (between Cu foil and electroless or DP), type M (in the mean area of electroless or DP) (see Figures 11.1 and 11.2). Figure 11.1 shows a SEM image for postseparation of type F. Figure 11.2 shows a SEM image for postseparation of type M. Both postseparations are related to electroless plating. More detailed qualitative evaluation of resin/DP, glass/DP, and copper/DP interfaces resulted from SEM photomicrographs. Figure 11.3 illustrates the interconnection appearance for a specific production site.

11.3
Electrical Resistance versus the Number of Cycles Classification Method

The correlation between the electrical resistance variations in DP interface as measured by IST testing technologies and the failure types in the whole interconnection was developed. The evaluation of DP quality is viewed as a process of pattern classification. The proposed approach is based on the use of similarities between vectors attached to each diagram electrical resistance versus time, for I/L-interconnect processing of IST minicoupons, as shown in Figure 11.4. Figure 11.4 shows electrical resistance pattern and the grid allowing pattern classification. Each vector corresponds to a diagram and this in turn to a tested sample. Matrices of similarity, algorithms, and grids of classification have been established.

Table 11.1 Microsectioning data: associated vectors first step of classification.

Number	Site	DP type	Cu thickness (mil)	Separation degree (internal data)	Classification (internal X-section)	Matrix (electrical data IST)	Classification, algorithm ($T = 0.9$)	Classification (independent laboratory)
1	5	Electroless	1.19	No	1	111000	1	1
2	15	Pd	1.0	Partial separation	0	111000	1	0
3	25	Conductive polymer	1.26	No	1	111000	1	1
4	4	Electroless	1.1	No	1	111000	1	1
5	6	Electroless	0.92	Separation F	0	100010	0	0
6	2	Electroless	0.96	Separation CM	0	100110	0	0
7	23	Nonformaldehyde	0.87	Separation	0	110001	0	0
8	21	Pd	1.18	Separation C	0	110001	0	0
9	14	Pd	1.18	Partial separation	0	111000	1	1
10	10	Graphite	1	No	1	111000	1	0
11	16	Pd	1.04	Separation C	1	000000	0	0
12	22	Pd	0.98	No	1	111000	1	0
13	8	Carbon	1.4	No	1	111000	1	1
14	20	Pd	1.14	No	1	111000	1	0
15	17	Pd	1.15	No	1	111000	1	1

1 mil = 25 μm.

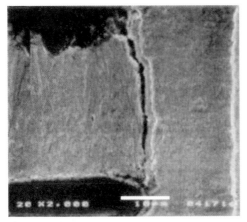

Figure 11.1 Postseparation *F*.

Figure 11.2 Postseparation *M*.

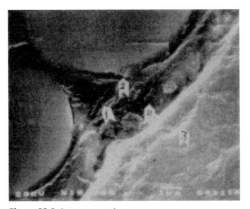

Figure 11.3 Interconnection.

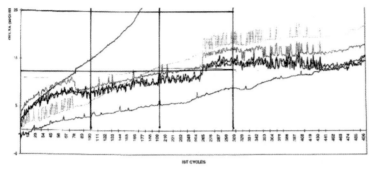

Figure 11.4 Resistance versus number of IST cycles.

11.4
Associated Vectors–Grids

To simplify the information, classification grids are used to examine the electrical resistance versus the number of cycle's data. The step length on axis is selected on account on available data and on the knowledge about the physics of the phenomenon. The approach is to start by considering the total range of parameter variation and to split this in few equal steps, usually two to four. Here, three steps on x-axis and two steps on y-axis are considered as a first trial, which results in a grid with six cells. The resistance versus cycles diagrams are transformed into vectors with six elements, $\langle x1, \dots, x6 \rangle$, that correspond to the six resulting cells, that is, rectangles. An example of the grid is shown in Figure 11.5. The step of 100 cycles is selected on x-axis, since the usual number of recorded IST cycles is from 300 to 500. The step of 25 mΩ is selected on y-axis since, usually a resistance increasing of 50 mΩ was recorded by this IST technology. The evaluation of resistance versus time, that is, cycles diagram is done in the next order: first, the three down cells in increasing order of time ($\times 1 \times 2 \times 3$), then, the three top cells in the same increasing order ($\times 4 \times 5 \times 6$). Each cell in the resistance versus number of IST cycles diagram will be marked "1" if it contains experimental data corresponding to that interval on x-axis and "0" otherwise.

If more intervals on the y-axis correspond to the same x-axis interval, due to sudden increase of resistance, only the higher resistance presence will be accounted for by the digit "1". Figure 11.6 contains illustrative examples. One examines not only the basic

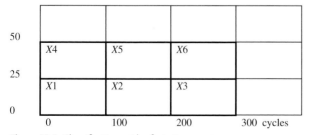

Figure 11.5 Classification grid – first step.

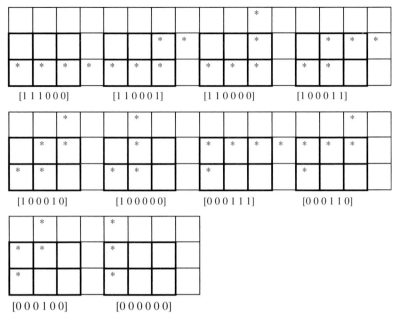

[1 1 1 0 0 0]   [1 1 0 0 0 1]   [1 1 0 0 0 0]   [1 0 0 0 1 1]

[1 0 0 0 1 0]   [1 0 0 0 0 0]   [0 0 0 1 1 1]   [0 0 0 1 1 0]

[0 0 0 1 0 0]   [0 0 0 0 0 0]

Figure 11.6 Examples of grids and associated vectors.

six cells of the grid but also six neighboring cells. The vectors associated with different experiment are of the type 1×6.

11.5
First Step of Classification

Table 11.1 presents examples of associated vectors resulting by applying the grid from Figure 11.5 to IST data. The classification algorithm is presented in what follows next (see also Chapter 5). The rule of classification is two samples or sites, that is, two vectors are assigned to the same class at the degree T, if their similarity is higher than T $(0 < T < 1)$. Obviously, various definitions of "similarity" and "distance" may be used. By using the classification algorithm, different partitions of sites in classes are obtained. Denote by $C(T)$ the resulting classification at the similarity degree T. It results $C(0.9) = \{\{4, 5, 8, 10, 14, 15, 17, 20, 22, 25\}; \{21, 23\}; \{2, 6\}; \{16\}\}$. This means that at the degree of similarity $T = 0.9$, the sites may be clustered in four quality classes: the highest quality class $\{4, 5, 8, 10, 14, 15, 17, 20, 22, 25\}$, the high-quality class $\{21, 23\}$, the middle-quality class $\{2, 6\}$, and the low-quality class $\{16\}$. An order of quality for the production sites is $\{4, 5, 8, 10, 14, 15, 17, 20, 22, 25\} > \{21, 23\} > \{2, 6\} > \{16\}$. The partition in just two quality classes could also be considered. Denote the high-quality and low-quality class by "1" and "0", respectively. In this case, at $T = 0.9$, class "1" $= \{4, 5, 8, 10, 14, 15, 17, 20, 22, 25\}$ corresponding to the highest quality; class "0" $= \{2, 6, 16, 21, 23\}$ corresponding to the lower quality. Table 11.1

includes also the results of the classification in just two classes for $T = 0.9$ based on associated vectors. The distance DD between different classifications are DD (algorithm, internal microsection) $= -81.76$; DD(algorithm, independent laboratory microsection) $= -67.5$. The DD was defined in Chapter 5. Relative distances have practical significance.

11.6
Second Step of Classification

If a more detailed partition in quality classes is necessary, a more detailed examination of resistance versus the number of cycle's diagrams can be performed. This will be restricted only to the higher quality samples having the associated vector [1 1 1 0 0 0]. The strategy is to select the best class and to restrict the next step of analysis to this class.

A new grid is proposed by splitting any down cells into two equal sized classes (see Figure 11.7). The classification results are presented in Table 11.2.

Applying again the classification algorithm, a more detailed order of quality for the sites results $\{4, 5, 8, 14, 17, 25\} > \{10, 22\} > \{15\} > \{20\} > \{21, 23\} > \{2, 6\} > \{16\}$. The last three classes $\{21, 23\}$, $\{2, 6\}$, and $\{16\}$ resulted in previous classification. The electrical resistance exam is limited to this two-level of splitting of the bottom cells as

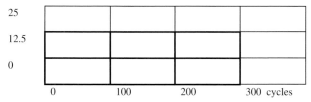

Figure 11.7 Classification grid – second step.

Table 11.2 Associated vectors – second step of classification.

Number	Panel sample	Site	DP type	I/L separation degree	Matrix (electric data IST)
1	1	5	Electroless	No	111000
2	2	15	Pd	Partial separation	100011
3	5	25	Conductive polymer	No	111000
4	11	4	Electroless	No	111000
5	38	14	Pd	Partial separation	111000
6	42	10	Graphite	No	110001
7	56	22	Pd	No	110001
8	64	8	Carbon	No	111000
9	71	20	Pd	No	000111
10	73	17	Pd	No	111000

in Figure 11.7 since a more detailed check up of resistance variations enters into the domain of electrical noise.

11.7
IST Lifetime and Electroplated Cu Thickness

For DP of high quality of interconnection, such as resulting from the second step of grid-based evaluation, it is possible to make use of data concerning IST lifetime and Cu thickness to characterize quality in more details. The failure mode for these IST minicoupons was the cracks in second copper. First, the interconnection evaluation and then a PTH evaluation is utilized, since the main objective is to evaluate interconnection and since PTH quality has a contribution to quality as resulting from statistical analysis. The ratio $Q =$ IST life time/Cu thickness was utilized as a new quality criterion. In this way, the higher quality class splits into three new subclasses: $\{4\} > \{5, 8, 17\} > \{14, 25\}$. It is assumed that for the first class $Q > 300$ (cycles/mil), for the second class $250 < Q < 300$ (cycles/mil), and for the third $200 < Q < 250$. Finally, it results in the classification of sites: $\{4\} > \{5, 8, 17\} > \{14, 25\} > \{10, 22\} > \{15\} > \{20\} > \{21, 23\} > \{2, 6\} > \{16\}$. The first three classes contain high-quality product from the point of view of interconnections. The last classification is based on the following strategy: use the Figure 11.5 grid, select the best and use the Figure 11.7 grid, and select the best and use the reliability index Q.

11.8
Summary

The type of DP is statistically irrelevant. The conventional statistical analysis does not allow discriminating between the types of DP. A detailed quality classification of sites according to the procedure proposed here and based on only the worst specimen for microsectioning of the same panel is $\{\{4\} > \{5, 8, 17\} > \{14, 25\}\} > \{10, 22\} > \{15\}$ $\{20\} > \{21, 23\} > \{2, 6\} > \{16\}$. The first three classes contain the higher quality product from the point of view of interconnection performance. These classes include {electroless, carbon, Pd, and conductive polymer}.

This confirms that for any type of DP, it is possible to find site-offering product of high quality. The result outlines the role of the site and technology implementation and shows that the tested DP technology is not enough robust to be generally implemented in PCB industry. The evaluation of DP quality without microsections, based on only electrical resistance measurements by IST, is significant. Here only in the third step of classification, the Cu thickness and IST life times are included.

The products of the above methodology are the grids, and the strategy of data evaluation (the grid order, the cell order in a grid, and the utilized degree of classification T). The method allows a drastic reduction of the experimental and analytical effort. The matrix columns in Tables 11.1 or 11.2 serve for new experiments classification. In this case, domains of the diagram electrical resistance versus the

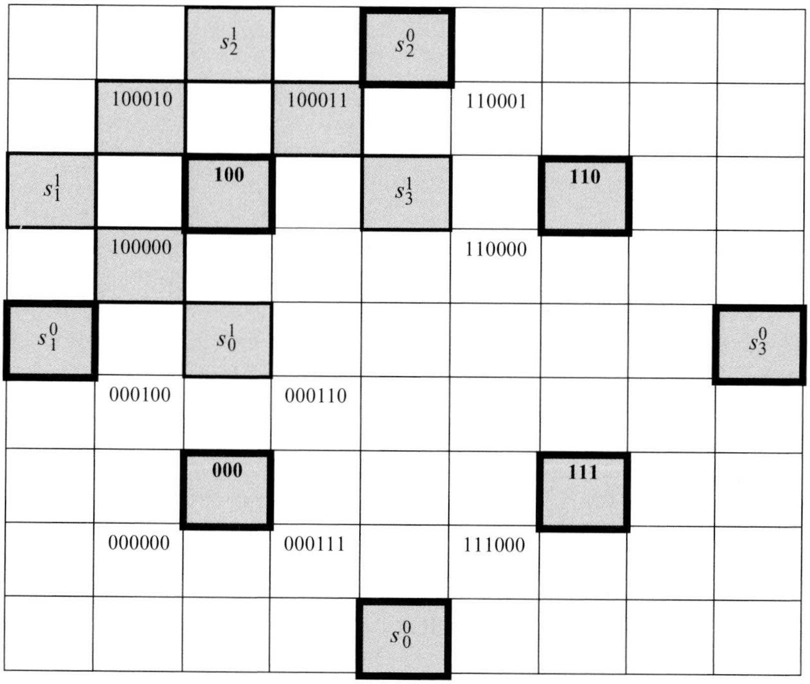

		s_2^1		s_2^0					
	100010		100011		110001				
s_1^1		100		s_3^1		110			
	100000				110000				
s_1^0		s_0^1						s_3^0	
	000100		000110						
		000				111			
	000000		000111		111000				
				s_0^0					

Figure 11.8 *SKUP* schema for direct plate.

number of cycles replace the DOE factors. The applied procedure can be viewed as a peculiar application of the general EDOE methodology. The classification steps are emphasized. The start is the central DOE with the matrix from Table 11.1. It contains 15 runs and six factors. One selects the best results and looks more into details, by forward steps. The new DOE is shown in Table 11.2 and contains 10 runs and six factors. Again, the best results are selected and a new criterion is introduced, namely, IST lifetime per mil copper thickness. The involved matrices are far from being the more informative ones but they correspond to practical conditions. New, nontested situations can be evaluated by backward classification steps.

11.9
SKUP Schema for Direct Plate

The data examination may be done in two separate steps of three digits. This corresponds to the digitalization, as shown in Tables 11.1 and 11.2.

Denote the three-digit vectors and six-digit vectors by $m=0$ and $m=1$, respectively. The *SKUP* elements are $S=(s^0, s^1)$, $K=(k^0, k^1)$, $U=(u^0, u^1)$, and $P=(p^0, p^1)$.

The schema includes at the first level $m=0$, the four vectors of three digits $0\,0\,0$, $1\,0\,0$, $1\,1\,0$, and $1\,1\,1$ and at the second level, $m=1$, the ten vectors of six digits related to previous ones and grouped as follows: $\{0\,0\,0\,0\,0\,0, 0\,0\,0\,1\,0\,0, 0\,0\,0\,1\,1\,0, 0\,0\,0\,1\,1\,1\}$,

$\{100000, 100010, 100011\}$, $\{110000, 110001\}$, and $\{111000\}$. The elements of K are expressed as discrete sets of digits representing the condition of DP choice. The states S are real valued and may describe the history of a DP search process.

Figure 11.8 illustrates this particular PSM. The three-digit vectors are the conditions at the level $m=0$. Let $000 = k_0^0$, $100 = k_1^0$, $110 = k_2^0$, and $111 = k_3^0$. The states and the conditions at the level $m=0$ are illustrated by high-thickness border cells. The system initial state is s_0^0. With possibility $p^0(k_0^0|s_0^0)$, the condition k_0^0 is selected. This is a three-digit vector symbolizing a specific DP behavior during the test. Based on this, the operator $u^0(k_0^0, s_0^0) = s_1^0$ describes the transition to the new state s_1^0 in the search process. Then, with possibility $p^0(k_1^0|s_1^0)$, the new condition k_1^0 arises. In the new condition, the operator $u^0(k_1^0, s_1^0) = s_2^0$ allows the search to reach the state s_2^0. With possibility $p^0(k_2^0|s_2^0)$, the condition k_2^0 is selected and so on. If several reasons impose $k_1^0 = 100$ as the accepted condition factor, as soon as the search arrives here, the analysis may be continued at the level $m=1$ for the test conditions $100000 = k_0^1$, $100010 = k_1^1$, and $100011 = k_2^1$. The transition from level $m=0$ to level $m=1$ is a forward step.

The system initial choice at the new level $m=1$ is s_0^1. The states and the conditions at the level $m=1$ are illustrated by medium-thickness border cells.

With possibility $p^1(k_0^1|s_0^1)$, the condition k_0^1 arises. This is a six-digit vector symbolizing a specific resistance pattern during the test. Based on this, the operator u^1, where $u^1(k_0^1, s_0^1) = s_1^1$, describes the transition to the new state s_1^1 of the search and so on. Each condition supposes the selection of other resistance patterns.

12
Plating Voids

12.1
Plating Voids Type

Plating "void" is the term used to describe discontinuities in PTH electroplated copper.

As technology advances, the ratio between board thickness and hole diameter becomes higher and the PCB difficulty, with respect to plating the holes, increases. Small hole voiding represents one of the common problems for most PCB manufacturers.

Figure 12.1 shows the appearance of the main type of plating voids as revealed by microsections. The pattern of void is important for failure root analysis and for corrective actions. These patterns allow limiting the number of factors in components DOE. This is critical, since several small DOE are more manageable than single big one. In fact EDOE replaces a big DOE with a network of small DOEs.

Table 12.1 shows the voiding failure mode and the processing steps that may be responsible for that type of voiding.

The process step contributors to different types of voids are marked by a star "*" in Table 12.1. There are twelve contributors that are typically lumped in six main groups each corresponding to a choice in the processing. The six groups are shown in Table 12.2.

Table 12.1 outlines two main types of voids associated with two main categories of processes. The process categories to be considered are before and after dry film.

Table 12.1 shows that plug hole void, drilling void, wedge void, electroless void, resin void, and glass void are basically rooted in the first class of factors including, material lamination, drilling, and electroless step. The other types of voids such as rim void, dry film plug void, dry film lock in void, electroplating void, etch-out void, and ring void are basically rooted in the second class of factors including dry film, electroplate copper, and etching. A matrix of the type W8,6,2 is useful in this case.

This matrix ranks the six factors in their potential significance order $F1 > F2 > \ldots$ $F6$. This corresponds to the natural order of manufacturing steps. The matrix contains the columns of Walsh–Paley matrix WP8 reranked in the order 4, 5, 6, 3, 2, and 7 that ensures more information than the other ranking of factors.

Evolvable Designs of Experiments: Applications for Circuits. Octavian Iordache
Copyright © 2009 WILEY-VCH Verlag GmbH & Co. KGaA, Weinheim
ISBN: 978-3-527-32424-8

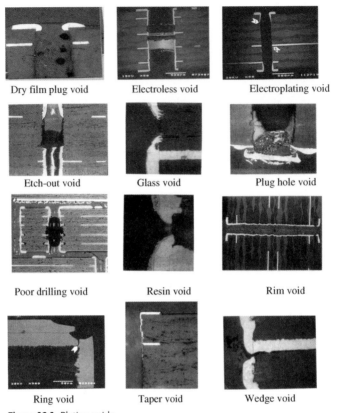

Dry film plug void   Electroless void   Electroplating void

Etch-out void   Glass void   Plug hole void

Poor drilling void   Resin void   Rim void

Ring void   Taper void   Wedge void

Figure 12.1 Plating voids.

The setting "−1" in Table 12.2 corresponds to nonmodified "old" parameters while "+1" corresponds to modified "new" parameters.

The significance of factors for the first design for grouped parameters at the level $m = 1$ is shown in Table 12.3.

Tables 12.4 and 12.5 show the level $m = 1$ of EDOE, when the large majority of factors are supplementary detailed.

This corresponds to a split DOE in which two different technologies or production sites are available.

For these factors, it is supposed that $F1 > F2 > F3 > F4 > F5 > F6$.

Tables 12.4 and 12.5 allow detailing the factors, for example, instead of considering the dry film as a single step, the exposure and the development are modified separately. This corresponds to forward steps. Backward steps are mediated by microsection optical evaluation and the return to the failure analysis guide from Table 12.1.

The factors $F1$, $F2$, and $F3$ will be maintained as before in the best variant, if the objective is to improve the process or in the worst variant, if the objective is to reproduce failure, to identify failures root and to select the conditions for failure.

Table 12.1 Plating voids failure analysis guide.

Failure mode	Mat	Laminat	Drill	Deburr	Desmear	EL	DF Exp	DF Dev	Surf preparation ECu	ECu	Etch-resist predip	Etch-resist plate
Plug hole	*											
Drill	*	*	*	*								
Wedge	*	*	*	*	*							
Electroless	*	*	*	*	*	*						
Resin	*	*	*	*	*	*						
Glass	*	*	*	*	*	*						
Taper	*	*	*	*	*	*						
Rim							*	*	*	*		*
DF plug							*	*	*	*		
DF lock-in							*	*	*			
Electroplating							*	*	*	*	*	*
Etch-out								*		*	*	*
Ring										*	*	*

Dev, development; DF, dry film; Exp, exposure; ECu, electroplated Cu; EL, electroless Cu; Mat, materials.

Table 12.2 Matrix W8,6,2.

#	F1	F2	F3	F4	F5	F6
1	1	1	1	1	1	1
2	1	−1	−1	1	1	−1
3	−1	1	1	−1	1	−1
4	−1	−1	−1	−1	1	1
5	−1	1	−1	1	−1	1
6	−1	−1	1	1	−1	−1
7	1	1	−1	−1	−1	−1
8	1	−1	1	−1	−1	1

Table 12.3 Grouped factors, $F1$ to $F6$, $m = 0$.

Factor	"−1"	"1"
$F1$-Materials and lamination, ML	Old	New
$F2$-Drilling, DR	Old	New
$F3$-Electroless, EL	Old	New
$F4$-Dry film, DF	Old	New
$F5$-Electroplated copper, ECu	Old	New
$F6$-Etching, ET	Old	New

Table 12.4 Detailed factors, before dry film $F1$, $F2$, $F3$, $m = 1$.

Factor	"−1"	"1"
$F11$-Material	Old	New
$F12$-Lamination	Old	New
$F21$-Drilling	Old	New
$F22$-Deburring	Old	New
$F23$-Desmear	Old	New
$F3$-Electroless	Old	New

Table 12.5 Detailed factors, after dry film $F4$, $F5$, $F6$, $m = 1$.

Factor	"−1"	"1"
$F41$-Dry film exposure	Old	New
$F42$-Dry film development	Old	New
$F51$-Surface preparation/Cu electroplate	Old	New
$F52$-Cu electroplate	Old	New
$F61$-Etching resist predip	Old	New
$F62$-Etching resist plate	Old	New

12.2
The *SKUP* Schema for Plating Voids

The failure analysis may be represented in a *SKUP* schema as shown in Figure 12.2.

The first design with lumped parameters is represented by trajectories of high-thickness border cells. The corresponding conditions at the level $m = 0$ are k_j^0, $0 \leq j \leq 5$. The six considered conditions are ML $= k_0^0$, DR $= k_1^0$, EL $= k_2^0$, DF $= k_3^0$, ECu $= k_4^0$, and ET $= k_5^0$.

The conditions are $F11 = k_0^1$, $F12 = k_1^1$, $F21 = k_2^1$, $F22 = k_3^1$, $F23 = k_4^1$, and $EL = k_5^1$ for the first detailed DOE at the level $m = 1$ and $F41 = k_6^1$, $F42 = k_7^1$, $F51 = k_8^1$, $F52 = k_9^1$, $F61 = k_{10}^1$, and $F62 = k_{11}^1$ for the second detailed DOE at the level $m = 1$.

				s_3^0		s_7^1		
							F41	
s_5^1		EL		s_6^1		DF		s_8^1
	F23						F42	
s_2^0		s_4^1				s_9^1		s_4^0
	F22						F51	
s_3^1		DR				ECu		s_{10}^1
	F21						F52	
		s_2^1		s_1^0 s_5^0		s_{11}^1		
	F12						F61	
s_1^1		ML				ET		s_{12}^1
	F11						F62	
s_0^0		s_0^1				s_{13}^1		s_6^0

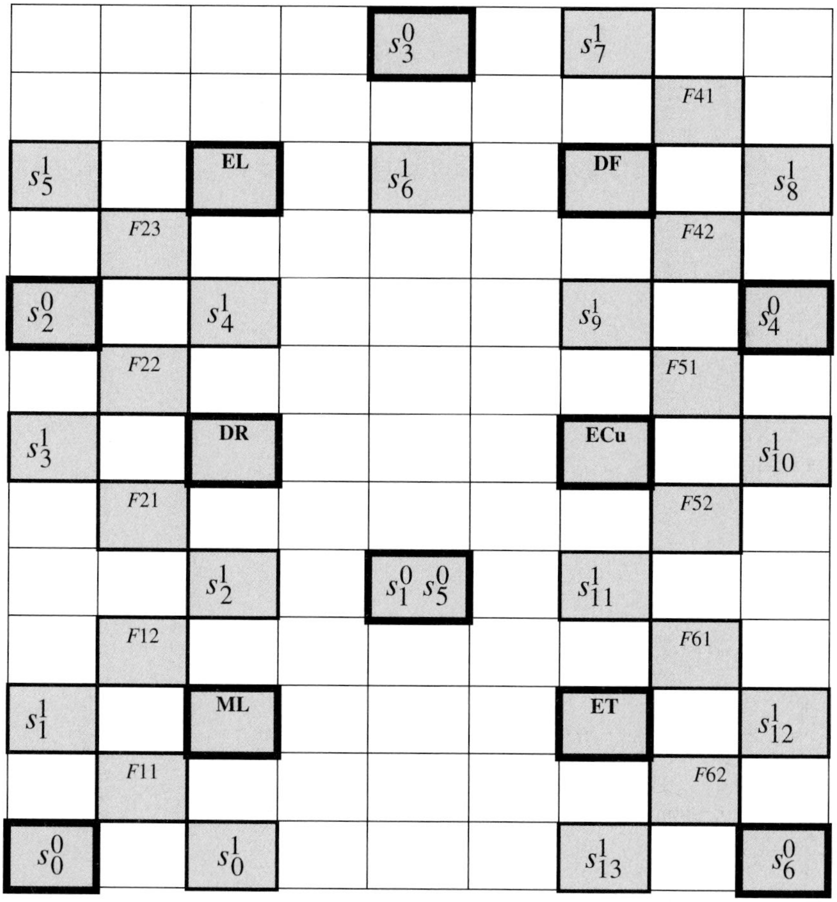

Figure 12.2 *SKUP* schema for plating voids.

Table 12.6 Sample of trouble-shooting guide.

Type of void pattern	Related process operation or material	Possible cause	Points to check and improve	Comments
Plug	Material	Material physical properties, degree of curing internal delamination	Pretesting and evaluate materials	Fast cure materials are more brittle
Total or partial plug, asymmetrical, and random location.	Drilling	Drilling debris	Check drilling parameters versus T_g. Check backup and entry materials. Check stack	Setup of drill configuration versus undercut and speed
To be looked in vertical cross section and horizontal section (sequential)	Drilling	Debris agglomeration, that is, conglomerate of resin, reinforcement material, and Cu foil, usually plated with EL and Electrolytic Cu	Check for debris after drilling–deburring	
	Drilling	DF solid residues entrapped due to rough drilling	Review drilling condition	
	Deburring	Entrapped brush material, burrs	Use brushes with water. Use appropriate brushes material	Some brushes materials are brittle
	Hole clean	Hole clean not appropriate	Use the flow: high-pressure rinse, first step with alternative jets on both sides. Ultrasonic-module in rinse. Medium pressure rinse, second step. Fresh water rinse, cascade. Dry	Alkaline cleaner with ultrasonic beneficial To pass two times through deburring–cleaning modules, second time deactivated brushes
		Drilling backup panel debris or entry material residues	Check back panel material and lubricant sheet	

EL Cu surface preparation	Not adequate for debris dislodging	Control vibration and surface tension in key steps
EL Cu	EL Cu flakes	Check for plating out, sleeves on heater, racks, and rollers
EL Cu	Particulates in sweller and PM	Check filtration and PM
EL Cu	Different particles from inside equipment, dirty rollers, etc.	Check line cleanliness and improve PM
DF scrubber	Entrapped scrubbing material (Al_2O_3, pumice) associated or not with Cu flakes	Use appropriate grit and rinses
DF lamination	Dirty tacky rollers	Clean and do PM
Cu plating solution	Flakes or solids of different provenience in Cu plating solution	Improve filtration
	Solid residues in solutions	Check filtration
Diverse processes environment or manipulation		Check anodic bags
		Best practice

DF, dry film; EL, electroless Cu; PM, permanganate; T_g, glass transition temperature.

The electroless step is considered as a condition at both levels, $m = 0$, that is, k_2^0, and $m = 1$, that is, k_5^1. The states are s_j^0, $0 \le j \le 6$ for the first detailed design and s_j^0, $7 \le j \le 13$ for the second detailed design. The trajectories at the level $m = 1$ are illustrated by medium-thickness border cells. They correspond to incomplete loops.

Observe that $K = ((k_0^0, k_1^0 \dots), (k_0^1, k_1^1 \dots))$ are processing conditions indicating the step.

The states are $S = ((s_0^0, s_1^0, \dots), (s_0^1 s_1^1, \dots))$. They may be examined step by step or only finally. The operators are $U = (u^0, u^1, u^2, \dots)$, while the possibilities are $P = (p^0, p^1, p^2 \dots)$.

The left side of the Figure 12.2 contains the first detailed DOE, while the right side contains the second detailed DOE. The initial application of the grouped DOE may suggest some dominant factors as, for instance, electroplate copper *F5*. In that case, for the next step of experiment, the factors *F5* and close to this, namely, *F4* and *F6* will be tested in detail, as suggested by the DOE matrix shown in Table 12.5. The result of implementation of the second step detailed design DOE should be compared to the void pattern shown in microsections. Definitely, there are mixed patterns as taper voids and there are patterns that have not been covered by Figure 12.1.

The DI experiment should select the main contributors in component designs and repeat experiments. If the factors involved in voiding and the void pattern are consistent, the corrective actions as in the example shown in Table 12.6 should be implemented.

Notice that often the root cause is not a single source, but several interacting sources.

Consequently, it is erroneous to assume that all plating voids can be corrected with a single change in the manufacturing process.

Part Four
Evolvability

Evolvable Designs of Experiments: Applications for Circuits. Octavian Iordache
Copyright © 2009 WILEY-VCH Verlag GmbH & Co. KGaA, Weinheim
ISBN: 978-3-527-32424-8

13
Self-Adaptive Circuits

13.1
Evolvability Levels

EDOE methodology highlighted new paradigms emerging in the study of complex systems. This refers to self-adaptability, proactiveness, and evolvability as discussed in Chapter 2.

Self-adaptability is the ability to perform tasks specific to biosystems, such as compensating a stress, self-building, self-repairing, self-organization, and so on. Self-adaptability is still a learning process and may be considered only as initial step toward evolvability.

The proactiveness is the ability to anticipate by design the new constraints before their effects manifest and ensure the strategy to manage these constraints. It is based on multiple scale modeling, prediction, and simulation capability and includes self-adaptivity.

Fully evolvable systems should be autonomous and creative, able to interact, to self-improve, and to take control of their changing environment to make opportunistic and advantageous use of unprogrammed events. Evolvability implies cognitive models embodied in real systems or devices. Proactiveness and evolvability corresponds to situations going beyond adaptive learning. The difference between them consists in the degree of achievement only. The ratio prebuilt part to self-building part is much higher for proactive circuits than for fully evolvable circuits.

In categorical terms, the proactive circuits correspond to a biased start in circuit development in favor of the U functor, whereas the evolvable circuits start from a more equilibrated situation between the U functor and the dual or adjoint constructions of P functor.

From their beginning, the circuits have been inspired by living cognitive systems such as the brain, since the printed circuits have conductive circuitry embedded in a dielectric nonconductive mass. But to achieve the level of biomimetic or cognitive tasks, the circuits should add new features such as memory, data acquisition and compression, learning by generating trials, self-analysis of incomplete quantitative and qualitative data, and finally creative evolvability.

Evolvable Designs of Experiments: Applications for Circuits. Octavian Iordache
Copyright © 2009 WILEY-VCH Verlag GmbH & Co. KGaA, Weinheim
ISBN: 978-3-527-32424-8

13.2
Self-Adaptivity

The self-adaptive circuits (SACs) are intended to have a degree of autonomy in new situations. SACs are almost fixed circuits. Elements of self-adaptability are naturally or intentionally associated with complex circuits. Existing adaptability trends may be accentuated by design and activated as opportunities.

The high sensitivity to initial conditions is a characteristic of any complex systems. This sensitivity means that for such systems apparently insignificant change of initial conditions may induce large changes of results. The high sensitivity is responsible for product variability and the failure of conventional control methodologies for complex systems. For many years, the circuit's complexity was considered as an enemy of the quality. Partisans of the robust design methods strongly promoted this idea. The claimed objective of robust design methodology was to find and keep a combination of product parameter values that gives reproducibility, that is, the smallest variation in the value of the quality characteristic around the desired target value. This may be possible for simple, mechanical, and quasilinear systems, but definitely it is a misleading task for high-complexity nonlinear systems. Challenging the robust design, the EDOE methodology suggests to accept unavoidable variability and use the opportunities offered by variability and complexity for circuit evolvability. Variability is not only inevitable but also essential for evolvability. The challenge is not to decrease indefinitely the variability but to establish when the illusory strict control of parameters by statistical process control (SPC), the geometrical or mechanical unrealistic precision, should be replaced with self-adaptive and evolvable designs, processing, control, and specifications.

13.3
Self-Adaptive Designs and Constructions

The multiple plated layers for PTH represent an example of self-adaptive constructions for reliability. The less reliable layer breaks when PTH is mechanically stressed. This allows more flexibility and redistributes the stress for the remaining test to be done among the resisting layers (Figure 13.1). It is a method to self-accommodate the failure. Figure 13.1 shows the three metallic layers' status during the stress test. The PTH will operate even if one of the plated layers fails. For accurate intermediary copper surface preparation, a multiple layer performs better than the single layer of the same total thickness. The multiple layers are recommended as design option. The order in which different layers have been applied is significant.

Self-repairing boards represent another well-studied case of self-adaptability that may be imposed from design. Large thermal excursions could induce PTH partial cracking. Solder may enter through barrel cracks, increasing in this way the PTH lifetime (see Figure 13.2). Figure 13.2 shows the failure pattern of a soldered hole. The self-repairing board is designed to carry extra mass that adapts to repair damage

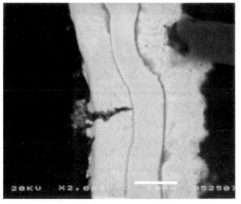

Figure 13.1 Multiple layers.

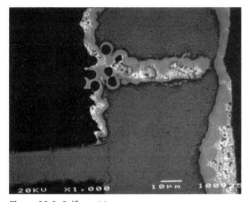

Figure 13.2 Self-repairing.

in functioning. The maximum heating temperature versus the solder eutectic temperature relationship is important in this case.

Open problems pertain to establishing the quantity, the location, and the properties of self-repairing material so that the repairing fluid reaches the failure sites developed randomly through the PTH.

Other examples of self-adaptive designs for reliable electronic devices are compensation devices (temperature compensations, encapsulations, and buffers), alternatives way to avoid failures (ensuring circuitry redundancy), multiple heat sinks with various domains of operation, and so on.

13.4
Self-Adaptive Materials

There exists a limited and controllable variability protecting the product from the high-risk inflexibility in unprogrammed conditions. Materials may have a self-adaptive behavior in different field situations. Eventually, they may outline multiple domains of properties allowing global versatility. Examples of self-adaptive or "active" materials are numerous: shape memory materials and piezoelectric (ceramic or polymers), ferroelectric, magnetostrictive, electrorheologic, and electrochromic materials, optical fibers, tunable dielectrics, electroactive polymers, conducting polymers, liquid crystals, and so on (Horowitz and Needes, 1997; Miller and Downing, 2002). The well-known tin–lead solder as well as some lead free solders outlines self-adaptive behavior.

An interesting application refers to multiple phase transition temperatures, T_g, for resin systems. Self-adaptability significant effects have been detected in the case of more T_g or for a domain of high T_g replacing the one-jump T_g transition. The lower T_g corresponds to transitions that induce preliminary copper cracking during thermal stress. The transition in steps allows a better distribution of failure along the plated through-hole. The derivative of heat flux in TMA measurements exhibits the multiple T_g phenomena (Figure 13.3). Figure 13.3 shows the derivative of heat flow as a function of temperature. T_g corresponds to maximums in such diagrams.

Multiple T_g constructions are obtained by blending laminate or by using the same laminate with different curing degrees.

Laminates with fillers and reinforcements represent another class of self-adaptive materials in circuits industry. The elements of self-adaptive materials rearrange, rather than deform or break locally, when mechanically stressed (Figures 13.4 and 13.5).

Figure 13.4 shows glass fibers, more light than the resin in electronic microscopy micrograph. The resin shows brighter inclusions of filler. Figure 13.5 shows only the resin and the filler playing this time for reinforcement.

Another element of plated copper adaptability is its crystallization or annealing. Selecting materials that experience annealing or recrystallization during their damage cycles makes the accumulation of damage much slower. Annealing is due to structure change within material during the application of external thermal or

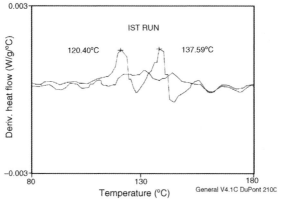

Figure 13.3 Multiple glass transitions.

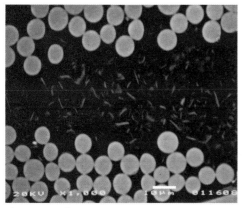

Figure 13.4 Filler.

mechanical energy. Even when no phase transformation occurs, the properties of some materials change with temperature. For the annealing period, the electrical resistance decreases as shown in Figure 13.6. Figure 13.6 shows IST electrical resistance as a function of the number of cycles. It suggests that instead of being a fatigue contributor, the first part of fatigue cycling increases the lifetime.

13.5
Self-Adaptive Processing

Self-adaptive material processing is an emerging control concept for the processing of materials. It seeks to integrate advanced sensors, quantitative process models, and process heuristics within a workstation environment to facilitate online control of product microstructure and properties (Upton, 1992). The adaptive material processing was formulated to address a set of processes in materials processing that do not

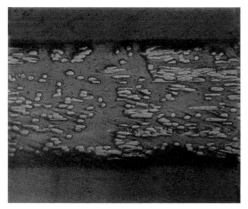

Figure 13.5 Reinforcement.

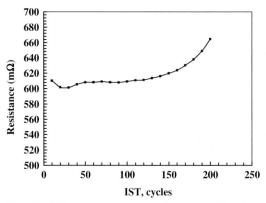

Figure 13.6 Electrical resistance as a function of IST cycles.

have adequate mathematical or physical models, which are subject to many stochastic factors acting at different scales and cannot be equipped by a complete set of sensors. The complex processing problem may find a solution in the frame of EDOE. The self-adaptive technology should have adaptability at different supplies. The final product may be repeatedly obtained despite fluctuating and variable feed. The starting challenge in the implementation of the adaptive system is how to build the knowledge-based system and how to acquire knowledge. The knowledge may be acquired using an online expert system that facilitates learning from proper expertise based on trial experiments performed according to the structure of an online classification algorithm. Self-learning and memorization are based on connecting DOE matrices. The system can learn the "best" regimes by fast classification methods. Ultrametric methods may be used to quickly automatically identify resistance patterns, shapes, substances, and different other features (Rammal, Toulouse and Virasoro, 1986; Iordache *et al.*, 1993a, 1993c). Vectors describing the species properties in a hierarchical order are actuated to generate relations of similarity. The intended system first generates and then changes by adaptive procedures the proper flow sheet and operations. The impact of having an adaptive technology is that it will be functional in unknown conditions. An example is the implementation of adaptive processing in lamination process when *in situ* sensors control the degree of cure.

Adaptive processing methods are also encountered in electroplating for circuits. It is known that plating at constant parameters is not possible for complex circuits. Plating is done at variable current density. Floating shields have to be used to compensate the board unavoidable variability. Selective plating and selective soldering are steps toward self-adaptive processing.

13.6
Self-Adaptability to Testing and Field Conditions

Expected behavior of tested samples during stress testing is the self-adaptability to test conditions. Obviously, the test conditions are related to field conditions.

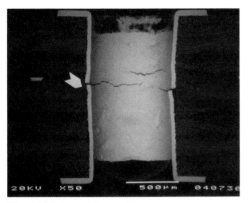

Figure 13.7 Incomplete cracks.

Adaptability of PTH barrel to failure is an example. In high thermal ramp situations, in which crack propagation is very fast, a tendency exists among some holes to produce noncircular cracks where the crack tips do not meet (Figure 13.7). Figure 13.7 illustrates an incomplete PTH crack developed after long-time stressing.

The noncircular crack is not a test failure not being detectable during electrical test and offering a high accommodation to thermal stress. This suggests that it is possible to guide the cracking by performing a spiral profile on the hole wall. This is a method of self-allocation of the failure area. The sequencing of tests is another method able to modify reliability. There are situations in which instead of damage accumulation some level of preliminary mild stressing proves to be beneficial for reliability and to increase the lifetime during the main reliability test. Submitting circuits to known amount of stress in conventional testers reveals unexpected correlations between damage and self-adaptability.

Baking for stress relieving is usual in circuit fabrication. A self-adaptive test or reliability control modifies the testing strategy according to the actual results. Table 13.1 summarizes certain characteristics of the self-adaptive testing in the complexity domain. The left side corresponds to the natural trend for simplicity and

Table 13.1 Testing for complexity.

Simple testing	Self-adaptive testing	Exhaustive testing
Available tests only	Test that customer really needs	All the specified tests
Test only one feature	Significant to quality features	All specified features
Discriminate quality	Flexible classification, sampling	Indiscriminating
Threshold of acceptance	Versatile criteria	Till failure
Fast (shock)	Self-adaptive, versatile	Slow (low fatigue)
Fixed domain, specific	Critical domain, less specific	Complete domain
Sequential	Multiscale, structured test	Parallel
One failure type	Failure mode, critical to quality	All known failures

Table 13.2 Self-adaptability as synergy and complementarity.

Quick, simple	Self-adaptability	Complete
Homogeneous	Self-structured	Hierarchical
Disordered	Self-organizing, coherent	Ordered
Nonconformity	Functionality	Total conformity
Variability	Self-adaptability	Invariability
Differentiating	Self-connecting net	Integrating
Dissociate	Systemic	Associate
Community	Synergy, complementary	Autonomy
Changeable	Elasticity, versatility	Robust, unchangeable
Adaptive	Autonomous	Innovative
Operational control	Multiscaled control	Strategy, planning
Sensitive	Integrated sensor/actuator	Actualize, driver
Variable tasks	Multipurposes	Fixed task, mission
Excitability	Active, multireactivity	Immobility
Simultaneous, fast	Multiscaled	Slow
Unspecified	Multiproperty	High specificity
Without memory	Multiscaled memory	Complete memory
Linear logic rules	Adaptive–predictive models	Axiomatic rules
Discontinuous	Self-structured net	Continuous
Local	Network	Global
Data driven	Data and goal driven, cognitive	Goal driven
Random, stochastic	Self-adaptive, evolvable	Determinist

quick solutions. This is the selected preference for short or easy ways to an objective according to the trend of least effort. The right side corresponds to another natural trend toward perfect or complete solutions. Both these ideal trends are extremes. The mean column expected to correspond to self-adaptability is not intended to be an average but a complementarity and synergy between the two extremes.

Table 13.2 presents self-adaptability as a synergy between the two dichotomies. The left side corresponds to the trend for simple and quick solutions. The right side corresponds to the opposite trend toward complete solutions. Self-adaptive systems show the reality of these pairings in that the elements of a dichotomy are combined in different ways.

14
Proactive Circuits

14.1
Proactiveness for Circuits

Proactiveness is a forward-looking perspective from engineering design point of view, enabling the circuit to modify faster in anticipation of the future constraints of the environment.

Proactive circuits (PACs) are circuits having the capability to change the preemptively embedded circuitry elements to keep on and to accomplish unprogrammed tasks. PAC makes use of self-construction elements offered by the basic generic frame and the environment. PAC is expected to outline object-oriented behavior. The evolutionary devices and sensors developed by the cybernetician Gordon Pask (Cariani, 1989, 1993), the "evolved radio" described by Bird and Layzell (2002), some developments of "evolvable hardware" (Thompson, Layzell and Zebulum, 1999; Mange et al., 2004) may be regarded as proactive circuitry implementations.

One way to achieve circuit autonomy is to have sensors constructed by the system itself instead of sensors specified by the designer. Cariani refers to "Pask's Ear" as a first example of such constructivist circuits. The Pask's system is an electrochemical device consisting of a set of platinum electrodes in an aqueous ferrous sulfate/sulfuric acid solution. When current is fed through the electrodes, iron dendrites tend to grow between the electrodes. If no or low current passes through a thread, it dissolves back into the acidic solutions. The threads that follow the path of maximum current develop the best. In the complex growth and decay of threads, the system mimics a living system that responds to reward that is more current and penalty that is less current. The system itself is able to discover the most favorable forms for the condition, which may embed information concerning other factors of the environment such as magnetic fields, auditory vibrations, and temperature cycles. This circuit was trained to discriminate between 50 and 100 Hz tones by rewarding structures whose conductivity covaried in some way with an environmental perturbation. The Pask's device, which may be considered an analogous realization of the *SKUP* quadruple and of the categorical frame, created a set of sensory distinctions that it did not previously have, proving that emergence of new relevance

Evolvable Designs of Experiments: Applications for Circuits. Octavian Iordache
Copyright © 2009 WILEY-VCH Verlag GmbH & Co. KGaA, Weinheim
ISBN: 978-3-527-32424-8

criteria and new circuits is possible in devices. The thread structures forming in malleable materials correspond to the category *S*, the resistance, capacitance, or ionic resistance linkage to energy is linked to the category *K*. The evaluation of the signal network developed in malleable material is part of the possibilities *P*. Amplifying servomechanisms may be linked to the operators *U*.

Following similar ideas, Bird and Layzell (2002) built an "evolved radio". Like Pask's ear, the evolved radio determined the nature of its relation to the environment and the knowledge of a part of the environment.

Bird and Layzell emphasized that novel sensors are constructed when the device itself rather than the experimenter determines which of the very large number of environmental perturbations act as useful stimuli.

The relation with von Uexküll *Umwelt* concept is obvious. Both of these devices, the Pask's ear and the evolved radio, show epistemic autonomy, that is, they alter their relationship with the environment depending on whether a particular configuration generates rewarded behavior.

PAC systems include the four basic parts of the Uexküll functional cycle: object, sensors, the command generator, and the actuator (von Uexküll, 1928). These parts are associated with the *SKUP* with the states *S*, the possibilities *P*, the conditions *K*, and the operators *U*, respectively. In the categorical frame, the perception is associated with the functor *P* and the action is associated with the functor *U*. They can be viewed as adjoint connections of the different categories *K* and *S*.

14.2
Evolutionary Hardware

Research for PAC, is associated with the domain of MEMSs (micro-electro-mechanical systems), MECSs (micro-energy-chemical systems) and to evolutionary hardware (Thompson, 1998; Teuscher *et al.*, 2003; Mange *et al.*, 2004; Mahalik, 2005).

MEMSs represent the integration of mechanical elements, sensors, actuators, and electronic circuits on a substrate through the utilization of microfabrication technology.

MECSs focus on energy transfer, microfluids, and chemical reactions. The PAC is the candidate for the "brain" part of evolvable systems, while MEMS or MECS would allow the microsystems to sense and control the environment.

In the associated *SKUP* quadruple, the environment corresponds to the states *S*, the PAC to conditions *K*, the MEMS or MECS control part to the operators *U*, and MEMS or MECS sense part to possibilities *P*.

For embedded EDOE, the MEMS or MECS and, ultimately, the PCB may be the physical support material. Coupling PAC with PCB and MEMS or MECS may ensure robustness and autonomy (Cheung *et al.*, 1997; Mahalik, 2005).

Evolutionary hardware represents the emerging field that applies evolutionary computations to automate adaptation of reconfigurable and polymorphic structures such as electronic systems and robots (Thompson, 1998; Thompson, Layzell and Zebulum, 1999).

Evolutionary computation methods in designs that take the performance of a scheme as prediction for the performance of a modified scheme are suitable for PAC development (Koza, 1999). Koza elaborated genetic programs that could design bandpass filters that are electrical circuits, which are able to separate signals of different frequencies. There is no explicit procedure for conventional designing of these circuits due to the large number of optimization criteria. The algorithms work by starting with simple circuits and evolving them. The program then created different variations, tested them, select the best, and used them for the next generations. Implemented on silicon, such programs may result in a circuitry that has attributes of novelty. The program may be used to generate PAC schemes.

An interesting suggestion for evolutionary hardware architecture is the Cell-Matrix (Macias, 1999). CellMatrix develops self-modifying, self-assembling, and self-organizing circuits. These circuits are designed for, and constructed out of, a unique hardware substrate. The CellMatrix may modify circuit architecture in the direction of locally connected, reconfigurable hardware meshes that merge processing and memory.

Immunotronics and embryonics are bioinspired concepts that have been successfully implemented in evolutionary hardware. Progresses have been reported in the construction of multicellular self-replicating systems (Mange *et al.*, 2004; Sekanina, 2004). This is significant since one of the characteristics of evolvability is self-reproduction. Mange and coworkers proposed the "Tom thumb" algorithms that made possible to design self-replicating loops with universal construction and computation that can be easily embeddable into silicon. For bioinspired projects, the elements of the *SKUP* quadruple are easily identifiable (Pattee, 1995). The symbolic description is linked to *K*, that is, the genotype and the constructed machine is linked to *S*, that is, the phenotype. The evolvability needs the closure ensured by possibilities *P*, corresponding to the ribonucleic acid, RNA-replicase in biosystems and related to the self-replication process. The description is translated via a constructor linked to the operators *U* corresponding to the ribotype, for example, transfer-RNAs. The multiscale structure in embryonics project is correlated to the four levels of organization analogous to molecules, cells, organisms, and population. The molecular level is represented by the basic field programmable gate array (FPGA) elements.

14.3
Electrochemical Filament Circuits

Based on electrochemical filaments development, a new class of proactive circuits, electrochemical filament circuits (ECFCs) became possible.

ECFC's construction starts with a generic frame representing the elements of the set of conditions *K*.

The *K*-frame elements may be those generated by the wave equation (WE). The process in *K* generates successive *K*-frames, *K*0, *K*1, . . . *Km* at different conditioning levels.

The generic circuitry represented by *K*-frames is completed by additional circuitry, traces, dendrites, filaments, supplementary matter, and corrosion or degradation products. The processing for these additional circuits is an *S*-process. *S* denotes the physical circuit based on filaments, threads, or microchannels for fluids allowing the electrical contact or interaction. The *K*-steps and the real environment *S*-steps have complementary contributions in circuit building. ECFC are expected to be at least partially autonomous. The autonomy includes the capability in building, assembly, modifying, organizing, repairing, and destroying. As a difference, if compared to self-adaptive devices based mainly on feedback, ECFCs make use of the preemptively embedded *K* frames. The appropriate *K* designs and the selective addition and the subtraction of appropriate elements from the environment are the processes allowing both self-functionality and proactiveness.

14.3.1
ECFC Design

Suggestions of proactiveness may be detected in conventional multipurpose circuit designs. These circuits have only holes and conductors, which have to be connected or interrupted according to the specific assembly needs. Often components are assembled directly on the component side so that it is possible to make small changes and keep the whole circuit under control. This technique permits easy adjustment and trials of different components to modify the circuit from the design stage. Adjacent to the multipurpose design methodology is the existing design reuse. The need to decrease time to market calls for the use of known good subcircuits or blocks as building elements. The subcircuits may be developed individually as component DOE in a design-centered EDOE.

The basic elements of ECFC technology are the *K*-valued generic frame, linked to the class of tasks, the environment media for self-construction in nonstationary or oscillatory fields, and the self-learning capability by exposure to environmental complexity and to variable tasks.

The ECFC that results from coupling the electrochemical filaments, ECF of different orders m (ECFm), with a preexisting *K*-frames $K0$, $K1$, $K2$, ..., Km, is considered here. The circuit may be described using the categorical tensor "*" that links different levels in circuitry: ECFC $= K0^* K1^* \ldots {}^* Km^* \text{ECF}0^* \text{ECF}2^* \ldots {}^*\text{ECF}m$

The tensor product "*" may be the categorical product "×" or the coproduct "∪".

The *K*-frame should be a quasicomplete printed circuit, with several opens. These opens allow the ECFC versatility and multiple potentialities. The environment is able to fill the opens sequentially in a way that ensures functionality. Potential geometrical variants and architectures for ECFC are dots, cells, hexagons, triangles, squares, circle arrays, circular crowns, dyadic structure, labyrinths and mazes, high-density circuitry, self-similar nested structures, tiles, and prefractals used for evolving antennas.

It was established that WE is able to generate fractal structures using categorical product "×". For example, Hadamard–Sylvester matrices reduce to Sierpinski triangles if only the "1s" are considered, whereas the "−1s" or with other notations, the "0s," are neglected, since they break the circuit (Barnsley, 1993).

The switch from categorical product to categorical coproduct determines the size and the shape of the circuit. The switch is determined by the oscillatory fields that accompany the ECFC development. The phenomenon is similar to the growing of tiny gold wire circuits in voltage-controlled colloids (Miller and Downing, 2002).

14.3.2
Materials for ECFC

The materials should offer opportunities for wet chemistry and solid physics to play significantly. ECFCs use composites and multiphase media. The materials should be as rich as possible in structural possibilities, for example, in phase transitions, on the edge of chaos and in nonlinear regimes. Interesting options are the existing self-adaptive or smart materials that allow phase transition, such as the piezoelectric, thermoelectric, electrorheological, electroactive polymers, and so on.

Laminates known as filaments nonresistant such as polyester rigid woven glass, paper phenol, or specially contaminated laminates represents valid opportunities, since they allow the fast formation of the electrochemical filament.

Possible *K*-frames conductor lines make use of materials such as Cu, Ag, Sn, Sn/Pb, Zn, Al, Mg, and Fe. Metallic inorganic salts for conduction may be sulfates, chlorides, or nitrates of Fe, Cu, Ag, Sn, Pd, Pt, Zn, Al, Mg, and catalysts. Metallic oxides may be useful as potential dielectrics. Damaged or fatigued, printed or integrated circuits represent new potentialities for proactiveness.

14.3.3
ECFC Processing

ECFC should be processed in the environment, that is, in real field conditions in which the circuit should be functional, such as

- Mechanical (vibration, pressure)
- TRB (temperature, relative humidity, bias) with direct current, alternating current, or pulse plating of variable frequency
- Light, radiation
- Cyclical operation of various types
- Superposed oscillatory fields.

These kinds of fields are the usual field of stresses for reliability tests. This suggests that PAC may result from some overtested circuits still able to show new capabilities.

A flow chart for ECFC fabrication is based on the following steps:

- Build *K*-frames based on the first-order WE solutions
- Select appropriate environment
- Introduce the *K*-frames and media in field conditions and allow periodic signals, stress field
- Develop the first level of filaments ECF0 during training for signal that needs to be sensed or for any encountered new signal

- Build the circuit ECFC = $K0^*$ECF0
- Repeat ECF0 procedure and allow ECF1 corresponding to another signal and so on
- Build the circuit ECFC = $K0^* K1^* \ldots * Km^*$ECF1*ECF2$^* \ldots *$ECFm
- Resulting circuits may be covered with gel, organic coating, photocoating, or lacquer to ensure protection and robustness.

The operators U from the associated *SKUP* describe the circuit PAC at different levels of its construction.

The ECFC would be a circuit useful and stable in its building conditions. For any new level, another frequency domain of oscillatory field is associated. As much as the oscillatory field still exists, the new level would be developed. If the structured dendrite structures were located in a specific field, the resulting structure would be able to recognize the patterns of that field. Learning and removing information is possible if any dendrite may continually be formed, broken, and regenerated. Training to discriminate signals may be accomplished with the help of WH or Haar waves. The similarity associated with WH, as defined in Chapter 5, represents the *P*-potentialities of the *SKUP*.

14.3.4
Potential ECFC Products and Applications

Potential applications in medicine are the proactive prosthesis, dosage of medicine based on biorhythms, drug testing, and chemical structure recognition devices. In the domain of defense/security/intelligence, one may consider self-configuring proactive sensors, proactive detection of chemical and biological patterns, sensitive bricks or clothes, and embedded pattern recognition devices that are able to be self-configured in areas denied.

For space applications, one considers the unmanned automatic circuits, and self-wiring or self-diagnosing circuits for critical systems.

In market study, one considers self-constructed neural networks for market examination based on historical data and on intermittent spiking phenomena.

Warnings in ECFC development, to mention, refer to

- Reproducibility
- Circuit growth process may be too slow and unmanageable for practice
- Risks of uncontrolled evolution, unpredictability desired and undesired dendrites
- Reliability problems, easy to damage in new media
- Long-term functionality, aging effects, and unknown failure mode
- Noise and chaos.

15
Evolvable Circuits

15.1
Evolvability Challenges

The EDOE was presented as a new modeling and simulation frame for complex problem solving. In addition, EDOE may support the cognitive architecture for evolvable circuit's (EC's) practical implementation. The challenge is to build circuits that take advantage and control of their environment in increasingly complex ways. EC is supposed to be an embodied EDOE able to run EDOE intrinsically, with emergent behavior.

ECs and PACs are closely related. The difference between PACs and ECs correspond to the degree of achievement from the point of view of evolvability. This is correlated to the degree of embodiment and to the scale. Transition from fixed circuits to evolvable circuits implies a change of scale. PAC refers mainly to micro- and mesostructured circuitry components while EC focuses mainly on molecular and nanomolecular structures. PAC is based on largely extrinsic designed and built circuits, whereas EC is expected to self-construct and modify most part of its circuitry based on a genotype–phenotype-like scheme inherent to evolvability and EDOE. PAC design is mainly from exterior whereas EC should be self-programmed from the interior of the devices in a complex interaction with its environment. There is a threshold below which PAC tends to become a fixed circuit and above which PAC may progress to become a full EC.

Unconventional principles, design configurations, materials, fabrication methods, testing, and applications have to be evaluated for evolvable circuitry (Abelson *et. al.*, 2000; Bedau *et al.*, 2000; Miller and Downing, 2002; Rasmussen *et al.*, 2004; Mahalik, 2005; Zauner, 2005).

Cellular automata suggest interesting architectures for EC software. An example is the EvoCA cellular automata system (Taylor, 2002). ECs are supposed to be organizationally closed for matter but informationally open. To realize evolvable systems, an important representational distinction should be between genotype and phenotype as well as a biotic structure. As illustrated by EvoCA, semiotically closed constructions may lead to "novelty." In EvoCA, the environment is represented by a layer made of cellular automata, the physical or dynamical part S, and the genotypes are repre-

Evolvable Designs of Experiments: Applications for Circuits. Octavian Iordache
Copyright © 2009 WILEY-VCH Verlag GmbH & Co. KGaA, Weinheim
ISBN: 978-3-527-32424-8

sented by a second genome layer, the inert or symbolic part K. Each genotype controls a given cell in the first layer and evolves through a genetic algorithm. EvoCA-like constructions lead to operationally closed evolvable circuits, embedded in a dynamic environment, having a metabolism-like potential and are capable of self-replication and self-maintenance.

15.2
Molecular Electronics

Molecular and nanomolecular systems represent the promising domain able to ensure the objectives of the EC, that is, to add evolution capability to devices, to self-construct systems going beyond learning, and to be able to act completely autonomous in an indeterminate environment. The circuits may be electronic, optical, molecular, microfluidic, and so on.

Critical for the EC autonomy is the embodiment of the computing capacity.

There are several molecular computation studies suggesting how to design synthetic chemical or biochemical circuitry able to perform specified algorithms.

A method of using molecules in computing architectures is reproduction of computer solid-state components with molecular structure. This is the usual approach taken in molecular electronics research. Its typical objectives are molecular wires, rectifiers, or transistors (Siegmund, Heine and Schulmeyer, 1990). Another research direction is the chemical computing based on the fact that chemical reaction networks are able to process information in parallel. Kuhnert, Algadze and Krinsky (1989) demonstrated contour detection, contrast enhancement, and same image processing operations on images projected onto a thin layer of a light-sensitive variant of chemical wave reaction medium. This system is a chemical realization of an associative memory and suggests the potential to implement learning networks by chemical means. The research into parallel chemical information processors led to artificial neural network (NN) design based on mass-coupled chemical reactors (Hjelmfelt, Schneider and Ross, 1993).

Studies in molecular and supramolecular design and engineering opened the prospects for the realization of electronic, ionic, and photonic circuits (Lehn, 2004). Orchestrated, supramolecular architectures deliberately designed to carry information allow to accelerate and to direct molecular interactions.

"Artificial chemistry" suggests innovative ways to go beyond the chemical kinetic level and encompass supramolecular interactions (Banzhaf, Dittrich and Rauhe, 1996; Dittrich, Ziegler and Banzhaf, 2001).

As expected, biomolecules provided potential substrates to build technical information processing systems as ECs. Biologically available conjugated polymers, such as carotene, can conduct electricity and can be assembled into circuits.

Among the biomolecules, the bacteriorhodopsin (BR) and the deoxyribonucleic acid (DNA) received extensive attention. Hybrid systems that combine the best features of biomolecular architecture, with optic, electronic, and microfluidic circuits, represent a necessary step in EC development. The hybrid character refers to both formal models and practical devices.

These hybrids are digital–analogue devices. The analogue aspects are related to rate-dependent processes and the digital aspects are related to macrostates and to macrostate transition rules. The issue of digital–analogue or symbolic-connectionist complementarity is closely related to the semiotic closure concept and evolvability for devices (Pattee, 1995; Cariani, 2001).

Some authors claimed that connectionist models as neural networks did not support symbolic processing and were incapable of adequately representing evolving cognitive structures (Fodor and Pylyshyn, 1988). Hybrid symbolic-connectionist models that may be implemented as hybrid devices allow a rebuttal of Fodor and Pylyshyn criticism in both theory and practice. The potential of the hybrid devices and hybrid models remains to be developed, but by all indications such representational method can provide strategies of unifying low-level connectionist and high-level symbolic models of cognitive processes.

15.3
Bacteriorhodopsin for Optoelectronic Circuitry

Early use of molecules in information processing has been in the field of optical computing. This suggested as candidate for EC base material the bacteriorhodopsin, which can serve as computer switch (Birge, 1995; Mann, 1996; Vsevolodov, 1998).

BR has two useful properties for molecular level calculation. It exhibits photochromic switching and shows photoelectric effect also.

The photocycle of BR, the sequence of structural changes induced by light, allows the storage of data in memory. Green, red, and blue light induce structural changes in BR. Green light transforms BR in an intermediate denoted by "k" that relaxes to the "o" state. Red light transforms the "o" state in "p" state that relaxes to "q" state. Blue light converts "q" state back to BR (Birge, 1995). Any long-lasting state can be assigned to digital values making possible to store information as a series of BR molecules in one or other state.

Discrete states such as "0", "1," and so on are functional for EC devices. With these identifications, the BR substrate may be the source for the symbolic language such as pixels and strings.

The photoelectric effect is another BR property useful for EC realization. Practical use of this property is exigent because it requires the preparation of BR films with highly oriented molecules. The possibility to interface BR electrically is the basis for several applications. The light of a specific wavelength range can be used to change the BR conformational state, and the conformation change is accompanied by a color change that can be detected by optical means. It should be observed that the circuits are in this case, at least in part, of optical type.

A significant step in the development of the optoelectronic circuitry and computing was the study of all-light-modulated transmission mechanism of BR films. When a yellow beam and a blue beam illuminate the BR film, the two transmitted beams suppress mutually. Based on this mechanism, in an all-optical operating device all 16 kinds of double-variable binary logic operations were implemented. The intensity of an incident yellow or blue beam acts as the input for the logic gate and the transmission bears the output of the gate. It is possible to turn this all-optical device into different states by using different wavelengths and different intensity illuminations.

Evolvability of hybrid symbolic-connectionist models and associated circuitry that may be based on the unique properties of BR will be evaluated in the following sections.

15.4
Embedded Symbolic-Connectionists Hybrids

Hybrid symbolic-connectionist techniques allow vectors representing the constituents of a symbol structure to be combined into a single vector representing the whole structure and for this vector to be decoded into the vectors representing the original constituents. In this manner, representations for compositional structures can be built up and then processed by NN.

An optoelectronic NN design, in which electronics is used to implement the summation, amplification, and threshold of the BR neurons, was described by Haronian and Lewis (1991). The interconnection strength for the neurons is given by the weights represented by the state distribution of BR molecules. Light sources can program the entire weight matrix in parallel. Microelectrodes coupled to the weight matrix feed the resulting photocurrent to amplifiers that have been selected by the threshold switches. A different wavelength of light flash restores the BR molecules that gave rise to a photocurrent.

The recursive autoassociative memory (RAAM) was among the NN models developed to address the question of how compositional structures may be stored within a connectionist framework (Pollack, 1990). The data for a RAAM network consist of a collection of trees and a representation that has a pattern of "0", "1," and so on for each terminal symbol occurring in those trees. The task for the network is to provide a means of compressing each tree into a representation, an activation vector, and reconstructing the tree from its representation. The elements of the *SKUP* quadruple are naturally associated with the RAAM elements. The input and output units may be associated with the set of the conditions *K*. Obviously, the discrete states "0", "1," and so on may be memorized by BR layers. The RAAM architecture contains encoding or compressor networks, associated with possibilities *P* in the *SKUP* frame. The RAAM also contains decoding or reconstruction networks associated with operators *U*. The hidden units are associated with the states *S*. As shown in Chapter 5, based on similarity calculations, classification trees may be associated with any matrices containing discrete information such as "0", "1," and so on. The RAAM has also basic structural elements in common with RIFS-random iterated function systems (Barnsley, 1993). The trees, embedded by representation into a "hypercube" of possible activation vectors, represent the transient states on the steps toward the attractor of the RIFS. The way a RAAM compresses the structure of the trees can be compared to fractal image compressing (Barnsley, 1993) that relying on the self-similarity of natural images is able to encode an image compactly by calculating and storing the RIFS maps that will decompress and recreate the image. The RIFS links to RAAM suggests that the RIFS embedding by laser stereolithography in BR may support a potential direction for evolvable optoelectronic circuitry.

Another promising NN architecture, in which conventional electronics may be adapted to molecular optoelectronics based on BR, is the distributed associative memory developed by Austin (1995). Associative memories operate differently from the memories typical for current computer architectures. This type of architecture takes a set of data often in the form of an image and scans the entire set of data in memory until it finds a set that matches it, as much as possible.

The architecture proposed by Austin (1995) uses for symbolic processing, the component-wise operations in GF(2). The categorical product "×" is in this case a vectorial outer product, whereas the categorical coproduct "∪" is a concatenation followed by superimposed coding. After coproduct implementation, it results in a correlation matrix memory (CMM). This offers a compact presentation of information. An advantage is that the proposed architecture is based on weightless NN. For BR systems as described by Birge (1995), the reading of "0", "1," and so on is the typical task. Light sources of different colors allow programming several matrices in parallel.

Another possible strategy to meet the challenges posed by connectionism critiques for both models and devices is offered by the so-called holographic reduced representations (HRRs) (Plate, 1995, 2000). Associative memories are conventionally used to represent sets of pairs of vectors. Plate describes a method for representing complex compositional structures in distributed representations. The method uses circular convolution to associate items that are represented by vectors. The representation of an association is a vector of the same dimensionality as the vectors that are associated. The method allows encoding relational structures in fixed width vector representation, but it should be noted that this increases the risk of missing the emergent structures. Plate calls his models holographic reduced representations, since convolution- and correlation-based memory mechanisms are closely related to holographic storage. The circular convolution may be associated with the categorical product "×", whereas the superposition may be associated with categorical coproduct "∪" (Plate, 2000). BR has potential for HRR practical implementation. Birge (1995) reports the development of an associative memory device that relies on the holographic properties of thin films of BR. These holograms allow multiple images to be stored in the same segment of memory, permitting large data sets to be analyzed simultaneously.

The properties of higher cognitive processes, and how they can be modeled by NN, have been extensively studied by Halford and collaborators (Wilson and Halford, 1994; Halford, Wilson and Phillips, 1998; Andrews, Birney and Halford, 2006). They proposed the so-called STAR model for analogical problem solving.

Assemblies of BR films in BR cubes may represent the appropriate material substrate for physical implementation of the STAR model. The BR molecule may be a node, that is, a neuron in the network. The rank of tensor used by Halford is linked to the arity of relation, that is, the number of attributes to the relation, and in the end to the Piaget stages of cognitive development. The STAR model uses a tensor of rank 3 to represent a predicate of two arguments.

Halford studies suggest that for early Piaget stages in cognitive development, the categorical coproduct "∪" prevails allowing the associative knowledge. This is a fast and parallel process. During the higher Piaget stages, the categorical product "×" seems preponderant, allowing the relational knowledge. It is a slow, sequential, effortful, higher cognitive process. The categorical product is naturally adapted to represent relations because its structure is analogous to the Cartesian product space in which relations are defined. The study of tensor product networks using distributed representations outlined the significant role of Hadamard matrices (Wilson and Halford, 1994). As shown in Chapter 4, these matrices are special solutions of the WE.

Notice that Halford and associates evaluated the significance of Klein-4 group and Latin squares for learning transfer in NN and cognitive systems (Andrews *et al.*, 2006). Such structures correspond to the INRC group studied by Piaget (Inhelder and Piaget, 1958) as well as to standard solutions of the WE model.

15.5
Temporal Synchrony for Embedded Symbolic-Connectionist Hybrids

A promising way of dealing with variable binding in connectionist systems is to use the temporal aspects of nodes or neurons activation. Phase synchronization can be used since it allows different phases in an activation cycle to represent different objects involved in reasoning and representing variable binding by the in-phase firing of nodes.

The implementation of temporal binding systems in BR support is naturally associated with the BR photocycles.

Based on temporal synchrony, SHRUTI system (Shastri and Ajjanagadde, 1993) provides a connectionist architecture performing reflexive reasoning. LISA system (Hummel and Holyoak, 1997; Hummel and Choplin, 2000) used the synchronous activation approach to model analogical inference. Both systems demonstrate that temporal synchrony in conjunction with structured neural representations suffices to support complex forms of relational information processing specific to cognitive systems.

The problem for such systems is their suitability for reflexive or reflective cognitive processes. Reflexive processes are linked to categorical coproduct, whereas reflective processes are linked to the categorical product.

While reflexive and reflective processes face different kinds of computational constraints, in most cases, the two types of processes interact and need to be integrated to perform a single task.

SHRUTI represents a restricted number of rules with multiple place predicates. There are several types of nodes or neurons in the architecture, denoted, for example, by circles, triangles, and pentagons that may be associated with different colors in BR-based films. Different types of nodes have different temporal patterns of activity. Light sources can program the temporal synchronization.

Relational structures such as frames and schemas are represented in SHRUTI by focal clusters of cells, and inference in SHRUTI corresponds to a transient propagation of rhythmic activity over such cell clusters. Dynamic bindings between roles and entities are represented within such a rhythmic activity by the synchronous firing of appropriate role and entity cells. Rules correspond to high-efficacy links between cell clusters, and long-term facts correspond to coincidence and coincidence-failure detector circuits.

SHRUTI was designed for reflexive reasoning tasks and the model is not suited to account for reflective processes.

To ensure applicability to complex situations, SHRUTI was coupled to systems activating the reflective component of problem solving. Such systems are capable of attention shifting, making and testing assumptions, and evaluating uncertainty. The resulting cognitive systems presented both reflexive and reflective capabilities and have been used to model decision making in imposed time frames.

LISA is a computational model based on temporal synchrony and designed for analogical inference and schemes induction. The data for LISA network consist of a collection of trees and a representation that has a pattern of "0", "1," and so on for each terminal symbol occurring in those trees. The tree contains a hierarchy of entities: problem, subproblems, roles, objectives, and semantics.

The task for the LISA network is to provide a means of compressing each tree into a representation, the so-called activation vector, and reconstructing the tree from its representation. The *SKUP* elements are naturally associated with the LISA elements. The problems to solve may be associated with the set of conditions K. LISA contains a driver network associated with operators U and the reflective reasoning. As a difference from SHRUTI, the initial LISA model was not developed to account for the reflexive processes. However, the representational structure of LISA provides at least a starting point for reflexive reasoning capabilities. LISA propositions are retrieved into memory via guided pattern matching. During retrieval and comparisons, the propositions are divided into two mutually exclusive sets: a driver and one or more recipients or receivers. The receiver network is associated with possibilities P. The switch between the reflexive and the refractive passes through the semantics. The LISA semantics elements are associated to the states S in $SKUP$.

The activation of semantic units is controlled by time. Often, the analysts do not have the time to allow runaway activation of semantics because they need to make inferences quickly. Notice that in contrast to reflexive inferences, which are fast, the reflective inferences may require more effort. An open problem is to establish, for imposed time frames, the number of switching from reflexive to refractive and the order in which the switching should be performed.

15.6
Embedded EDOE

The perspectives of a hybrid optoelectronic device based on the properties of BR molecules, in which conventional electronics is used to implement DOE analysis, are evaluated in the following.

Photocycle and photoelectric effects allow a direct writing DOE embedded in the BR-based substrate. BR memorizing digits should be complemented by standard electronics able to perform the real-valued operations.

As shown in Chapter 4, the DOE results as particular solutions of the wave equation. Consider the following solution:

$$Y(T, Z) = Z \oplus (V \otimes T) \tag{15.1}$$

A computing "cell" with three BR molecules is retained here for illustration purposes. Table 15.1 shows the wave equation solution for $V = 1$.

Table 15.1 is resulted from Galois field, GF(3), calculations and is a 3×3 Latin square. The factors are the time steps 0, 1, 2, the molecules #0, #1, #2, and the operations $0 = g$, $1 = r$, $2 = b$ corresponding to the three colors green, red, and blue able to induce transitions. The time is multiple of the same time step.

Standard DOE table may be developed by indicating the conditions associated with any element of the 3×3 Latin square (Table 15.2). Experimental results of DOE application may be the resolution or any other value or data to be memorized.

The DOE selects the significant results and also the significant factors by standard ANOVA calculations done by an external computer. This is Fourier analysis over the real field for the device functioning parameters.

Table 15.1 Convection model, $Y(T, Z)$.

$Z\backslash T$	0	1	2
#0	0	1	2
#1	1	2	0
#2	2	0	1

Table 15.2 DOE associated with three-molecule cell.

Experiment	Molecule	Time	Operation
1	#0	0	g
2	#0	1	r
3	#0	2	b
4	#1	0	r
5	#1	1	b
6	#1	2	g
7	#2	0	b
8	#2	1	g
9	#2	2	r

Successive steps will continue the experiment in the direction of beneficial results. The new experiment means a new DOE based on GF(m) algebra calculation and the wave equation. Following the EDOE suggestion, hardware may be achievable in 2D or 3D structures with concentric hierarchically located levels or planes. Light sources should be placed externally (Birge, 1995).

Based on special BR properties, new classes of evolvable circuits, embedding and evolving DOE, became possible. The evolvability for the proposed architectures is the challenged result. As far as the EDOE structures are concerned, after the implementation of the DOE matrix of the type Wn, m, s or Ln, m, s, it is required to perform at least two steps: factor evaluation, on columns in DOE, and experiment classification, on rows in DOE. It is necessary to define thresholds as degrees of acceptability for results. This helps to decide when to recognize a pattern to be classified as new. Various areas throughout the chosen EC layers may be written and addressed simultaneously. It is conceivable to embed Wn, m, s or Ln, m, s matrices in any active areas with memory. EC would be built by using in succession similar additive and subtractive steps as for printed circuits and integrated circuits fabrication. Matrices such as Wn, m, s or Ln, m, s play the role of masks in printed or integrated circuits fabrication. These evolvable circuits should be able to drive the input signal and to decode the signal in a manner similar to logical thinking processes. As a difference, if compared to conventional circuits, this kind of EC will be continuously formed and erased, allowing the operation to be in forward and backward succession. The parallel search may be organized to achieve amplification, resonance, and coherence. The EC works associatively as well as serially. By parallel processing, the experiments would be performed at once, and the recorded results can be presented simultaneously to the central DOE. The EC should be able to record data from different areas to analyze and give rise to a decision. This means that EC needs to have monitoring functions, that is, sensors, and executive functions, that is, actuators, since the long-term technological challenge is to get results by EC, independent of any external analyst or "operator." The EC should be a system that confronts the environment having the ability to adapt autonomously. New environmental conditions for EC may be materialized by a new row in the existing, embodied, component DOE matrices. This is the discrete symbolic step of the EC. Then follows the step in which real field values are associated with discrete DOE. This real-valued step goes after data expansion and precedes data compression. With a learned degree of acceptability, the sensor information goes backward and is classified in inner levels or layers and finally comes back to the center. In this way, the material embodiment may regenerate the symbolic description represented by DOE.

15.7
Hybrid Controlled Microfluidic Circuits

In microfluidic devices, the circuitry from printed or integrated circuits is replaced or completed by microchannels for fluids. The MEMS became in fact MECS (Mahalik, 2005). The transport of molecules in complex biological or chemical process may be programmed as the electric current in standard electronic circuits (van Noort, van Noort, Wagler and McCaskill, 2002; Verpoorte and de Rooij, 2003; Erickson and Li, 2004).

The microfluidic devices suppose the existence and the development of sensors, able to monitor changing environment, and of actuators able to influence environment, coupled with computing and control capabilities for communication and data processing, all physically wired together. Tangen *et al.* (2006) presented elements of an interesting development in this direction. It focuses on the application of online programmable microfluidic bioprocessing as a vehicle to the design of artificial cells.

The electronically controlled collection, separation, and channel transfer of the biomolecules are monitored by sensitive fluorescence setups. This makes combinatorial fluidic circuitry and biochemical reaction circuitry feasible.

The basic elements of the *SKUP* quadruple may be identified for the "biomolecular console" described by Tangen *et al.* (2006). The reconfigurable electronic interface is linked to the space of conditions *K*. The microfluidic network represents the states *S*. This includes chemicals' reservoirs and products. The parallel actuator network is related to operators *U*, while the monitoring system is linked to possibilities *P*. An electronic computer guides and controls the molecular circuits and ensures their cyclic functioning.

Another promising microfluidic technology consisting of a fluidic layer with a network of microchannels superposed on layer with external computer programmable electrodes and actuators controlling the flow has been proposed by Goranovic, Rasmussen and Nielsen (2006).

The basic elements of the *SKUP* are obvious for this technology. The genetic channel is linked to the space of conditions *K*. The temperature cycles ensure the gene replication. This fits with the cyclic character of the time *T*.

The metabolic channel is naturally linked to states *S*. The replication of selected protocells is linked to operators *U*, while the metabolism of selected protocell step is related to the possibilities *P*. The switch from categorical product to coproduct is determined by the oscillatory temperature fields and is able to control the protocell replication.

An important specificity of this microfluidic device is the realization of closed or loop operations, essential for the transition from fixed circuits to evolvable circuits.

15.8
Reconfigurable Microfluidic Circuits

A reconfigurable array of modular microfluidic circuits inspired from cyclic operations of separation is presented in the following discussion. In this case, the circuit scheme is associated with conditions *K*, while the circuit of fluids and molecules through this scheme is linked to the states *S*. This complementarity is at the root of computing potentialities.

The schema presented in Figure 15.1 corresponds to an array of four-state modules similar to square tiles. Figure 15.1 shows the elements in *K* only.

1	—	2		3	—	2
|		|		|		|
0	—	3	→	0	—	1
↑						↓
1	—	2		3	—	0
|		|		|		|
→ 0	—	3		2	—	1

Figure 15.1 Reconfigurable array of modules.

The four steps of any cell are 0, 1, 2, and 3. The 0123 rectangle or loop defines a single module. The coupling of modules may be flexible. After each step, the 0 for the next module may be reached from 1 or from 3 of neighboring modules. The steps 1 or 3 should embed sensors and actuators. They correspond to possibilities P and operators U as elements of the related $SKUP$ quadruple.

The stacked configuration of multiple cells results if the tensor product interpretation in WE solution is a coproduct "$\times$".

In the quest for evolvability, transition from $3 \to 0$ connection type to $1 \to 0$ connection type of two modules may be triggered by the presence of the answer of interest in steps 3 or 1. The output of any module can be configured to be driven by its output or by signal arriving from a central programming unit for 3 or 1 gate.

The overall behavioral effects coming from the scheme of the physical construction are implicitly taken into consideration. Only nearest-neighbor interconnections between modules were enabled in the scheme from Figure 15.1. Multiple levels of modules may be considered in 3D space. The 3D scheme organized as Latin cubes based on orthogonal Latin offers interesting suggestions for high compactness of circuitry (Hedayat, Sloane and Stufken, 1999).

For 2D or 3D scheme, recurrent connection paths through the scheme by which a module output can indirectly affect its own input are possible.

Evolution is allowed to exploit the capability of the scheme freely. The strongly interactive signals may be forced to explore their space of possible scheme and may create new unexpected patterns of interconnections and circuits.

The switch between product and coproduct interpretation allows involving more timescales in the circuit development. The resulting evolvable device is a hybrid automaton.

15.9
Self-Constructed Molecular Circuits and Computing

Self-construction and separation in classes may be considered as computational processes and may be utilized to build information processors. Observe that the basic elements of the *SKUP* quadruple are naturally associated with any self-construction or separation processes. Suppose that from an unstructured environment *S* some molecules considered as symbols are able to assemble in a supramolecular structure linked to the conditions space *K*. These *K* structures should be recognized by a receptor and possibly amplified to provide an action *U*, redirected toward the unstructured environment *S*. The selection of specific symbols from the environment is done according to possibilities *P*. This may be a process driven by an optimization criterion, for instance, energy or entropy production minimization or maximization (Prigogine, 1980; Dewar, 1993).

The self-construction may be described by WE, too. According to the interpretation of the tensor product, two main types of configurations result. The tree-like forms result if the tensor product is a categorical product, and a multiple cell-stacked configuration is to be expected if the tensorial product is a coproduct. The transition between the two configurations is mediated by environment conditions.

Elements of the general scheme of self-constructed computing are present in different DNA experiments (Adleman, 1994; Winfree, 2000).

Adleman proposed an approach for information processing with bioprocesses that allowed solving combinatorial problems by using a specific set of DNA molecules.

DNA-based computing consists of four basic operations: encoding, hybridization, ligation, and extraction. Problem solutions are obtained through an exhaustive parallel search by means of the pattern recognition intrinsic to DNA hybridization, that is, self-construction of the complementary DNA strands. Involved chemical reactions such as the activity of restriction enzymes, ligases, and polymerases, or simple hybridization can operate in parallel and this explains the possibility of solving complex problems.

Following similar ideas, cellular automata architectures describing DNA self-constructed circuit patterns for various forms of DNA tiles have been studied by Winfree (2000). Cook, Rothemund and Winfree (2004) showed how several common digital circuits, including demultiplexers, random access memory, and Walsh transforms, could be built in a bottom-up manner by using biologically inspired self-construction.

As established, the Walsh–Hadamard matrices are particular solutions of the wave equation. Table 15.3 shows a solution of the kinetic model in which we suppose the rate *Q* to be constant in the wave equation.

Table 15.3 Kinetic model, $Y(T)$.

$Q \backslash T$	000	001	010	011	100	101	110	111
000	1	1	1	1	1	1	1	1
001	1	−1	1	−1	1	−1	1	−1
010	1	1	−1	−1	1	1	−1	−1
011	1	−1	−1	1	1	−1	−1	1
100	1	1	1	1	−1	−1	−1	−1
101	1	−1	1	−1	−1	1	−1	1
110	1	1	−1	−1	−1	−1	1	1
111	1	−1	−1	1	−1	1	1	−1

It is in fact the so-called Hadamard–Sylvester matrix, similar to Sierpinski triangle presented by Cook *et al.* (2004). To highlight this parallelism, the bold letters are used for the wired "1" cells. It was assumed that "−1" breaks the circuitry. The nonwired digits are italicized. Notice that only two digits "0" (replaced here by "−1") and "1" need to be present in this case.

Based on operations in GF(4), Sierpinski square-like fractals may be generated (Carbone and Seeman, 2002a, 2002b).

At the present stage, a number of researchers are rather skeptical whether existing DNA-based computation strategies will ever follow bacteriorhodopsin on its path in information processing. Critical problems with DNA circuits and DNA biocomputers are related to their inflexibility and the ineffective accommodation of the variety of computation requests in real conditions. The assembly of DNA molecules of "tiles" has been designed to simulate the operation of any Turing machine. The self-construction DNA structures may be mapped naturally onto the grammars of the Chomsky hierarchy (Chomsky, 1966; Winfree, 2000). However, for strictly algorithmic operations, the DNA tiling computer cannot compete with silicon machines. It was expected that the DNA computers may be eventually advantageous for the complementary domain of computation, beyond Turing machine capability. Again, it is not the case for the self-assembled DNA circuits as much as they map the Chomsky hierarchy of grammars.

The 1D chain of DNA or the 2D crystal tiles represent only the informational, that is, the *K* part of the *SKUP*. Major parts of the actual DNA computation have been accomplished with human operator involvement. To solve complex problems, the *K* structure should be part of *SKUP* quadruple. *K* elements should be recognized by a receptor and amplified to provide an action *U* toward the external nonassembled environment *S*. The selection of specific symbols from the environment *S* would be done according to possibilities *P*. For this reason, the tiling needs to be flexible, and the tiles can be cycled through alternating assembly and disassembly stages. The self-construction and reconstruction operations may be programmed by using glued and unglued tiles (Carbone and Seeman, 2002b). According to the signification of the tensor product in WE solution, two main types of configurations result. The tree-like forms result if the tensor product is a categorical product, and a multiple tile-stacked configuration is to be expected if the tensorial product is a categorical coproduct. The switching from one tensor interpretation to another is induced by environment changes. Interactions between tiles and between tiles and their environment are mandatory to challenge Turing machines (Wiedermann and van Leeuwen, 2002). Notice that this Turing versus beyond Turing computation issue parallels the classic debate about language acquisition between Chomsky's preformationist views and Piaget's constructivist and emergentist views.

15.10
Genetic Code-Like Mechanism for Molecular Circuitry

The study of living cells offers more suggestions for artificial biocircuitry and embedded computation. Molecular-scale circuit fabrication using the principles of genetic code construction was presented in the study of amorphous computing.

The translation of genetic information in a living cell is based on the interaction between informational type of molecules, such as DNA and RNA, and the functional molecules such as amino acids. The genotype of cells is laid down in a linear sequence of four nucleotides: A (adenine), C (cytosine), U (uracil), and G (guanine). The four bases C, G, U, and A might form 64 different simple triplet patterns, that is, codons. The 21 amino acids and the start and stop signals are coded redundantly by these 64 codons.

The amino acid assignment depends on the order of importance of bases and their positions in codons.

The nucleotide hierarchical ordering for amino acids assignment purpose is $C > G > U > A$. The hierarchy was established starting from the observation that C, in position 2 in codons, is anytime able to be the source of a single amino acid. G in position 2 is able to determine the amino acids in majority of cases, U only in some cases, and A never. In other words, the C base is anytime a single message, whereas A is credited with at least double message. For G and U, the roles are mixed but G is able to pass stronger than U codification messages.

We may associate with any base in codons a two-digit vector (hydrogen bonds, chemical nature). The first digit refers to hydrogen bonds and the second to the chemical nature.

For codification purposes, the hydrogen bond is more significant than the chemical nature. We may denote by "1" the high number of hydrogen bonds for G and C and also by "1" the pyrimidines as the chemical nature for C and U.

Consequently, we will use "0" to denote the low number of hydrogen bonds for A and U and "0" for chemical nature purines for A and G.

In this way, we may associate with any base a two-digit vector as follows: C: (1,1), G: (1,0), U: (0,1), and A: (0,0). This corresponds to the hierarchy $C > G > U > A$ and to the real numbers 3, 2, 1, and 0 associated with C, G, U, and A, respectively.

There are some symmetry features in genetic code. These symmetries help in explaining regularities and periodicities as observed in proteins. The relevant algebraic group to describe the symmetries of the bases{C, G, U, A} should be a group of order 4. There are only two possibilities for group structure, the cyclic group C(4) and the Galois group associated with the Galois field GF(4) (Findley, Findley and McGlynn, 1982). The genetic code may result as a product of C(4) or GF(4) groups. The product construction appears as a method to obtain multilevel solutions of the wave equation.

The tensor product signification in WE solution is in this case the categorical product "×". The solutions correspond to different conditioning levels, that is, different steps for the time T in the development pathway for code. Here, we consider only two levels of development, the level $m = 0$ corresponding to the apparition of individual bases and the level $m = 1$ corresponding to doublets. The complete code contains triplets, involves the level $m = 2$, and offers a typical genetic code table.

For two-level cases, the elements of WE are $Y = y_0 y_1$, $T = t_0 t_1$, $Z = z_0 z_1$, and $F = f_0 f_1$.

Table 15.4 Single bases, y_0.

G	U
C	A

The wave equation reduces for $V = 1$ and $Q = 0$ to two similar equations, one for each level:

$$\frac{\partial y_0}{\partial t_0} \oplus \frac{\partial y_0}{\partial z_0} = 0 \tag{15.2}$$

$$y_0(0) = f_0 \tag{15.3}$$

$$\frac{\partial y_1}{\partial t_1} \oplus \frac{\partial y_1}{\partial z_1} = 0 \quad \text{and} \tag{15.4}$$

$$y_1(0) = f_1 \tag{15.5}$$

Suppose that the initial condition for two levels is

$$Y(Z, 0) = f_0 \times f_1 \tag{15.6}$$

Here "$\times$" denotes the categorical product.

The solution of the model 15.2–15.6 will be

$$Y(T) = y_0 \times y_1 \tag{15.7}$$

Moreover, if $f_0 = f_1$, then $y_0 = y_1$.

Let us consider a possible scenario for genetic code development from primitive bases.

The initial step of development of the genetic code-like frame contains the bases C, G, U, and A. This situation corresponds to the vector $y_0 = (C, G, U, A)$

A folding operation allows rewriting y_0, as shown in Table 15.4. This is a 2×2 matrix or table.

For the next level of development, each nucleotide C, G, U, and A may be the start for four classes of double bases.

Table 15.5 contains 16 doublets. The single bases C, G, U, and A are maintained in this table for illustrative purposes.

Table 15.5 Double bases, $y_0 \times y_1$.

GG		GU	UG		UU
	G			U	
GC		GA	UC		UA
CG		CU	AG		AU
	C			A	
CC		CA	AC		AA

Elements of the *SKUP* quadruple will be identified in the following discussion.

K is the symbolic description of the system, and in this case, the first and second bases in codons. *S* represents the realization of the doublet message, that is, the associated polypeptide chains or real molecular circuits. *U* describes the capability to pass from doublets or codons to the protein reality. This corresponds to translation in biological cells.

The capability of states *S* to reactivate and modify the symbolic description is characterized by the possibility *P*. *P* expresses the backward causation from the organism level such as the control of which genes are expressed. The backward control is not stored in memory but is part of the dynamical processes, and it is not heritable (Pattee, 2000).

The control is of hybrid type since the *SKUP* appears as a hybrid automaton.

Figure 15.2 shows the conditions set from Table 15.5, and also the resulting amino acids.

The amino acids coding may be associated with doublets. Strong doublets such as CC, CG, CU, GC, GG, GU, and UC codify for one acid only. The other doublets codify according to the third base in codons.

For easiness, only contiguous trajectories are examined here.

S, *K*, *U*, and *P* will be vectors denoted as follows: $S = (s^0, s^1)$, $K = (k^0, k^1)$, $U = (u^0, u^1)$, and $P = (p^0, p^1)$. Upper index refers to levels, whereas lower index refers to time steps.

The first level $m = 0$ of single bases is inactive for amino acids assignment.

Examples of notations are as follows: A $= k_0^0$, U $= k_1^0$, G $= k_2^0$, C $= k_3^0$ at the level $m = 0$; AA $= k_{00}^1$, AC $= k_{03}^1$, CU $= k_{31}^1$, GC $= k_{23}^1$, GG $= k_{22}^1$ at the level $m = 1$; and so on.

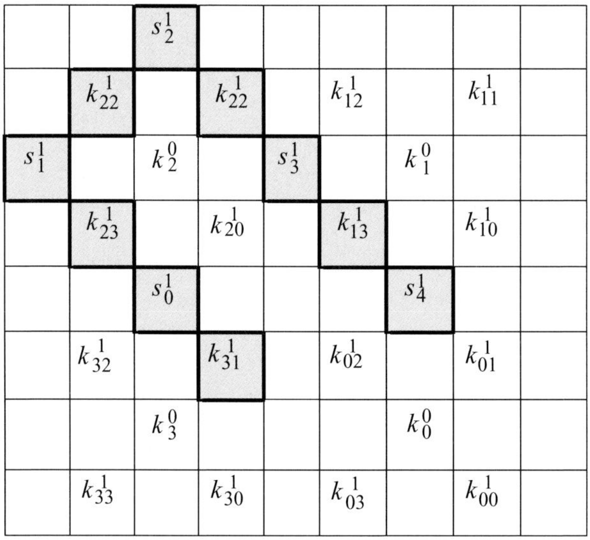

		s_2^1					
	k_{22}^1		k_{22}^1	k_{12}^1		k_{11}^1	
s_1^1		k_2^0		s_3^1		k_1^0	
	k_{23}^1		k_{20}^1		k_{13}^1		k_{10}^1
		s_0^1				s_4^1	
	k_{32}^1		k_{31}^1	k_{02}^1		k_{01}^1	
		k_3^0				k_0^0	
	k_{33}^1		k_{30}^1	k_{03}^1		k_{00}^1	

Figure 15.2 Doublets and polypeptides.

It is known that CU is coding for Leu. Correspondingly, the states at the second level, $m = 1$, will be $s_0^1 = $ Leu, $s_1^1 = $ Leu $\cdot$ *Ala*, $s_2^1 = $ Leu $\cdot$ Ala $\cdot$ Gly, $s_3^1 = $ Leu $\cdot$ Ala $\cdot$ Gly $\cdot$ Val, and $s_4^1 = $ Leu $\cdot$ Ala $\cdot$ Gly $\cdot$ Val $\cdot$ Ser.

The states at the level $m = 1$ are represented by medium-thickness cell borders in Figure 15.2. The operator u, associated with this one-level process, is of the type $u^1(s_0^1, k_{23}^1) = s_1^1$.

Here, s_1^1 is the amino acid associated with the codon k_{23}^1 coupled to previous chain of amino acids. Possibility, such as $p(k_{23}^1 | s_0^1)$, depends on the dynamics of amino acids.

Observe that $k_{31}^1, k_{23}^1, k_{22}^1, k_{22}^1, k_{13}^1$ is an informational circuit in the K space, whereas $s_0^1, s_1^1, s_2^1, s_3^1, s_4^1$ is a real circuit in S space.

The living cell genetic mechanism includes the following four parts: the amino acids and proteins, the DNA and RNA, the translation molecules such as aminoacyl-tRNA, and enzymes such as RNA replicase. In the $SKUP$ quadruple, they are correlated to real states S, conditions K, operators U, and possibilities P, respectively. The S-valued circuits represented by proteins are easily separable. Due to code ambiguity, they offer a nonunique solution to the problem encoded by K.

A molecular-scale circuit fabrication, using the above principles of genetic code-like construction, was already presented by Abelson *et al.* (2000). One of the ideas was to build scaffolding components of collagen. Collagen consists of an offset triple helix in which each strand has a sequence of amino acids of the form Gly·Pro-X-Gly·Pro-X-Gly·Pro-X . . .. In such sequences every third amino acid is glycine and many of the other amino acids are proline. The amino acids denoted by X can be arbitrary, other than the glycine and proline residues. The structure forms a rigid straight rod were the side chains of the X amino acids points outward. It is assumed that the Gly·Pro amino acids assure that the protein folds as rigid rods, and by choosing appropriate X amino acids, it is possible to control how these rods assemble into more complex structure. According to Abelson *et al.* (2000), the nanoscale circuit fabrication may begin by building a layout that incorporates the size of molecular electronic components and the ways that they might interact. For each collagen strut, which links only to its selected targets, the DNA sequence to make that kind of strut is assembled according to a procedure that inserts the DNA only into the appropriate location. For amorphous computing systems, the DNA sequences represent the conditions K in the associated $SKUP$ quadruple. The collagen and other sequences of proteins represent the space of states S linked to circuits. U and P are linked to the procedures describing K and S interactions. The presented genetic-like model is of interest for the new emerging discipline, synthetic biology, which studies the design and the construction of new biochemical systems, such as the genetic circuitry. Just as the engineers design electrical circuits based on known physical properties of materials and then fabricate functioning circuits and processors, the synthetic biologists design and build biological circuits. The research is motivated also by emerging technologies such as MEMS or MECS devices that are making it possible to bulk manufacture tiny computing elements integrated with sensors and actuators and embed these into smart materials and structures.

15.11
Conventional Circuits versus Evolvable Circuits

For the forthcoming PAC and EC fabrication, a natural query is why do not use traditional methods, such as the physical and chemical study, followed by modeling and extrinsic implementation of the models in the usual computer-based control of circuit fabrication. The answer is that the envisaged control and computing tasks are impossible to be extrinsically operated for evolvable systems of high complexity. In conventional circuit design, the majority of nonlinear interactions that could possibly contribute to the problem are deliberately excluded. The properties characterizing EC constructions should be, at least in part, the consequences of their own dynamics of the computational environment, not of the decision of the designer who is anyway unable to predict the evolution of its construction. ECs are supposed to work for their evolution more efficiently than an external computer or operator can do. ECs have the potential to be developed toward an autonomous system allowing survivability in completely unforeseen situations. It was observed that the more an autonomous system is optimized to perform a task, the less adaptivity it has to deal with unexpected changes in an uncertain environment. This implies that in complex environments evolvability rather than optimization may be an appropriate measure of a circuit's potential to carry out tasks. Complex systems, natural or artificial, seem to opt for evolvability rather than for optimization. This may be because in a complex environment it is impossible to achieve the optimum, particularly when there are strong interactions between conditions K and states S. One way to proceed is to diversify several acceptable circuit options in a given environment and to let them

Table 15.6 Comparison of conventional circuits and evolvable circuits.

Conventional circuits (PC)	Proactive or evolvable circuits (PAC, EC)
Single objective for any fabrication step	More general classes of objectives
Defined, based on previous learning	Undefined, open for learning, innovative
Top-down, linear	Top-down, bottom-up, cyclic, recursive
Aims for best solution, optimal	Makes workable, viable, active devices
Looks for perfect elements	Accepts elements with small defects
Conventional design, detailed models	Generic design, based on wave equation
Clear processing steps, complete data	Incomplete data and variable ad-lib steps
Independent of previous designs	Use everything at hand, if useful
Insulate the elements, serial or sequential	Combine elements, distributed, parallel
Builds	Builds, disbands, embeds, and reorganizes
Divide and conquer	Divide and integrate, opportunistic
Maintain functionality in different media	Sensitive to environment, multifunctional
Restricted, static	Less restricted, rich, dynamic
Isolate from medium protection	Medium, opportunistic exploitation
Avoid variability, interactions, transitions	Accept, use variability, interactions
Reliable	Robust, multireliable
High maintenance	Low and proactive maintenance
Catastrophic degradation	Degradation in steps, hindered

evolve. This means that evolvable circuits may have several possible nonoptimal but viable and useful architectures. This implies the discovery of environment properties that can be utilized to solve the imposed tasks.

Table 15.6 summarizes some of the differences between conventional circuits and unconventional ones, such as proactive circuits, PAC, and evolvable circuits.

At the present technological level, a manufacturing project grouping all the described faculties of EC is unrealistic, but it is expected that EC of increasing capability would be manufactured in small steps. Gradually, it is possible to make more and more evolvable circuits as the understanding and resources grow.

16
Evolvable Manufacturing Systems

16.1
Manufacturing Paradigms

In the beginning of the twentieth century, the concept of "mass manufacturing" characterized by the production of the same product in large scale was introduced. This paradigm was widely accepted and implemented, but in the last decades, it became evident that it could not respond to the challenges of a dynamic and worldwide manufacturing. The mass manufacturing systems became incapable to deal with variations in the type of product or to customize.

In the 1980s, several companies introduced, in practice, the related concepts of "just-in-time" and "lean manufacturing." Just-in-time concept implementation consists in having the right material at the right place and time, eliminating stocks. The lean manufacturing initiative aims to shorten the timeline between customer order and shipment by eliminating waste.

Just-in-time and lean manufacturing organizations invest heavily both in time and money to obtain facilities that are optimized for a particular task but deteriorates rapidly outside the design envelope due to unavoidable task variability.

"Computer-integrated manufacturing" paradigm of the 1980s consists in the integration of the enterprise activities, through the use of information technologies, which allows the exchange and sharing of data among the various enterprise units.

The main advantages were the increase of productivity, flexibility, quality, reduction of design time, and work in progress.

Confronting with complexity challenges, it became clear that monolithic, centralized, and static computer-integrated manufacturing approach does not fulfill modern functional requirements. In computer-integrated manufactures, a central machine computes schedule for the facility, dispatches actions through the factory in keeping with the schedule, monitors any deviation from plan, and dispatches corrective actions. This approach may impede flexibility since it is centralized; it relies on global plans and it precedes execution with complete planning and scheduling.

Evolvable Designs of Experiments: Applications for Circuits. Octavian Iordache
Copyright © 2009 WILEY-VCH Verlag GmbH & Co. KGaA, Weinheim
ISBN: 978-3-527-32424-8

In the 1990s, the "agile manufacturing" concept was introduced to challenge the lean manufacturing strategies. This concept refers to the ability to adapt quickly and profitably to continuous and unexpected changes in manufacturing environment. Agility is supposed to impact the entire manufacturing organization including product design, customer relations and logistics as well as production (Kidd, 1994). The underlying principles are

- Deliver value to the customer
- Ability to react to changes
- Value of human knowledge and skills
- Ability for virtual partnerships.

In agile manufacturing strategy, it is important to consider that human factors and organizational knowledge are equally important as the advanced knowledge is. The symbiosis between human factor and advanced knowledge is favored.

Agility paves the way for competitiveness in global markets. Manufacturers who can change volume of production rapidly can avoid sinking large amounts of capital in excess inventory, while still maximizing returns in periods of increased demand.

The current manufacturing is confronted with the market and product changes such as

- Total globalization
- Increasing production pace
- Significant decrease of production cycle times
- Migration from mass production to mass customization.

To challenge the market evolution, the product manufacturing systems of the future should be characterized by

- Direct response to market changes
- Sensitivity to customer needs
- Migration from central to distributed control
- Autonomous and cooperating production units.

Stochastic nonlinear phenomena, incomplete data and knowledge, combinatorial explosion of states, random change in environment, and the frame problem are some examples of difficulties encountered in contemporary manufacturing.

Such difficulties justify the need for continuous evolution of the manufacturing paradigms. Consequently, new manufacturing system concepts have been proposed and tested. Their roots can be found mainly in the study of biological cognitive sciences and social organization (Ueda *et al.*, 2001).

Examples of new manufacturing paradigms showing more or less evolvability potentialities are

- Fractal manufacturing
- Holonic manufacturing
- Biological manufacturing
- Virtual manufacturing
- Multiagent manufacturing.

16.2
Fractal Manufacturing System

The "fractal company" consists of independent self-similar nested units, the fractals, and may become a vital organism due to its organizational structure.

The fractal manufacturing uses the ideas of mathematical theory of complexity and from biology, the companies could be composed by small components or fractal objects that have the capacity to react and adapt quickly to the new environment changes. A fractal object is self-organized and self-similar. These are general attributes of evolvable systems. The breaking of fractal objects into other fractal objects has the particularity of generating objects possessing organizational structures and objectives similar to the original ones (Ryu, Son and Jung, 2003).

Main features of the implemented fractal manufacturing systems are

- Self-similar units, providing services
- Consistency of overall objective system through objective formation process
- Handling changes through self-organization and dynamic restructuring
- High individual dynamics of fractals
- Efficient information system.

The basic component of the fractal manufacturing system is the so-called basic fractal unit (BFU). It consists of several modules such as analyzer, organizer, resolver, and controller (Ryu, Son and Jung, 2003).

The basic elements of the $SKUP$ quadruple may be identified in BFU systems. As discussed, fractal configurations result from solutions of the wave equation (WE). These are elements of the space of conditions K. The top–down U and bottom–up P interactions allow hierarchical communication on multiple scales.

In BFU, the environment represents the space S. P is related to the chain sensor, observer, and analyzer. U is related to the chain actuator, controller, and organizer.

The observer, analyzer, organizer, controller, resolver, and the knowledge database may be considered as the elements of K. This means that K is highly structured in this case.

Each BFU provides services with an individual goal and acts independently, while resolving conflicts through cooperation and negotiation.

The BFU structure may be replicated at several levels in a self-similar fractal manner.

The cooperation between factory fractals is characterized by high individual dynamics and maximum ability to adapt to the influences of their respective manufacturing environments by evolvability.

16.3
Holonic Manufacturing System

The "holonic manufacturing system" is the paradigm that translates the "holon" concept developed for living systems and social organizations into a set of appropriate concepts for manufacturing industries.

The term "holon" describes the basic unit of organization in living organisms and social organizations. The holon can represent a physical or logical activity such as that of a machine on order or a human operator. The holon can be broken into several other holons, a procedure that allows the reduction of problem complexity. A manufacturing holon comprises a control part and an optional physical processing part. Multiple holons may dynamically aggregate into a single higher level holon. The holon concept may be correlated to concepts such as Piaget's schema and von Uexküll's functional cycle.

The "holarchy" is a system of holons that can cooperate to achieve an objective. The holonic manufacturing system is a holarchy that integrates the entire range of manufacturing activities.

The holonic manufacturing systems paradigm is part of the next generation of distributed control and introduces the hierarchical control within a heterarchical structure. This innovation makes available the combination of robustness against disturbances, presented in heterarchical control with the stability, and global performance optimization presented in hierarchical control. The implementation of this concept requires that decision power must be distributed between the central and local holons, that is, there exists a switch between hierarchical and heterarchical control. In categorical terms, this corresponds to a switch from product-to-coproduct constructions. The function of central holon is to advise the local holons. When disturbances occur, the autonomy of holons increases, whereas during normal functioning, the autonomy of local holons decreases and they follow the central holon's input as, for example, the scheduling plans.

The holonic manufacturing system design starts with a forward control step in which the definition of all the appropriate holons and their responsibility is established. Compared to traditional methodologies, a rather vague responsibility than a precise function for each holon is established. This facilitates the backward control. Complementary controls ensure system evolvability.

The random manufacturing system (Iwata, Onosato, and Koike, 1994) can be considered as a holonic type system. Its basic characteristics are as follows:

- Autonomous machine systems
- Dynamics machine grouping
- Task allocation
- Shop floor control by a reward and penalty system.

The information flows from the center that receives the order at the conditioning level $m = 0$, toward task master at the conditioning level $m = 1$, then toward task managers at the conditioning level $m = 2$, and finally toward the machine managers at the conditioning level $m = 3$.

Typically, the conditions K are represented by the manager's plans and the states S by the accomplished plans at different levels. The operators U describe the activity of task allocation at different levels. The communication, control, and reward activities are linked mainly to P possibilities.

16.4
Biologic Manufacturing System

The "biological manufacturing system" is a paradigm trying to adapt to the manufacturing world the concepts of biological nature as self-construction, self-organization, adaptation, and evolution. In biological manufacturing system, the autonomous entities are the manufacturing cells that have the capacity to self-organize in case of disturbances or change in demand.

The biological units are represented by genetic information found in DNA and adaptive information found in brain neurons. The biological manufacturing concept imposes that manufacturing units have associated information represented by a code and adaptive information.

The biological manufacturing system is developed under the ideas and concepts from biology and assumes that the manufacturing companies can be built upon open, autonomous, cooperative and adaptive units that can evolve. By analogy to biological cells, the manufacturing units have associated information represented by a code with static and adaptive information.

Biological manufacturing constituents may be naturally associated with the elements of the $SKUP$ and the biosystems. A minimal biological cell includes the following four parts: the amino acids and proteins, the DNA for genetic code, the translation molecules as aminoacyl-tRNA, and the enzymes as RNA replicase, for instance. These parts are correlated with S, K, U, and P, respectively. In the biological manufacturing system, S corresponds to material flow, K to information, U to operators or procedures, and P to coordinators.

It should be noted that the theoretical biology concepts represented the source of inspiration for manufacturing, starting from the engineering design (Yoshikawa, 1989; Tomiyama and Yoshikawa, 1987).

Modern engineering design is performed as an evolvable process. The so-called General Design Theory formulates engineering design as a process initiated by ideal specifications such as functional requirements in the function space. This corresponds to the space of conditions K, in the $SKUP$. The designer is able to match partial structural information in the attributes space corresponding to S. The operator U is linked to deduction stages that connect specifications to design solutions.

The possibilities P include the abduction inference to accomplish the closure of the design cycle. Abduction is the method to integrate knowledge that satisfies the two aspects of the creative design that creates new products and expands knowledge.

Most of the current computer-aided design (CAD) systems employ the hierarchical decomposition strategy, a form of analysis thought process corresponding to categorical product interpretation of tensor in the K model. Such a strategy can lead to refinement of the existing design, but does not always lead to new, creative designs. Moreover, it happens that the design becomes too large.

The switch from categorical product to coproduct controls the size of the search space and allows the emergence of new designs. Similar ideas have been studied as divergence/convergence strategies in engineering design. The divergence step, correlated to categorical product, implies understanding the problem and creating solutions. The convergence step, correlated to coproduct, selects solutions for further development.

16.5
Virtual Manufacturing System

Starting from the 1990s, the worldwide market composition forced some companies to develop and implement the "virtual enterprise" concept. This is a paradigm that refers to temporary alliances of enterprises that come together to share skills and resources to better respond to business opportunities and whose cooperation is supported by computer networks. The "extended enterprise" or "consortium" structures are particular cases where different enterprises are preponderant in the network organizations in a mature market or enterprises choose to cooperate in dynamic market conditions.

Despite having all attributes of an enterprise, the virtual enterprise would not be a permanent organization. Virtual enterprises comprehend multiple autonomous enterprises behaving globally as a single one. Each autonomous entity tends to focus its core competencies on outsourcing complementary product, expecting to reduce risks and costs while maximizing market opportunities. Organizations face a quasicyclic evolution process, adopting the most advantageous structure at different stages of the evolution.

The computer models allow simulating different aspects of the manufacturing, previously, without using the real facilities, and to accelerate and optimize the design and production of a manufacturing product in real manufacturing systems in this way.

This kind of enterprise may evolve in a dynamic environment, where a new market opportunity triggers the creation of a new enterprise structure providing the market with new products.

The need for improvement of enterprises is based on complementing, reactivity with awareness and partnership, flexibility with cooperation, autonomy with agility, and information with knowledge.

16.6
SKUP Schema for Virtual Manufacturing Systems

An evolvable manufacturing system represents an implementation of EDOE concepts in manufacture organization.

An example of evolvable manufacturing organization is presented in what follows next (see also Lucca, Sharda and Weiser, 2000). The associated *SKUP* schema linked to this virtual manufacturing is shown in Figure 16.1.

The center of frame is occupied by the objective, the product O.

The main steps in product realization are $O1$, cost evaluation; $O2$, process planning; $O3$, selection of manufacturing sites and task assignment; and $O4$, control.

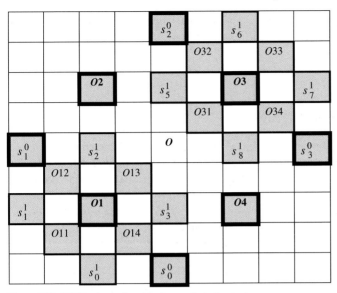

Figure 16.1 *SKUP* schema for virtual manufacturing systems.

For cost estimation, the subfactors are O11, locate similar product; O12, determine step costs; O13, contract description cost; O14, bid price; and so on. For planning, the subfactors are O21, locate similar products; O22, determine remaining processes; and so on.

For task assignment, the subfactors are O31, select potential cooperating man-ufactures; O32, process assignment; O33, offer to cooperating manufactures; and O34, manufacturing.

For control tracking, the subfactors are O41, select suppliers; O42, track processes; O43, assessment; and so on.

The diagram shown in Figure 16.1 illustrates the linked *SKUP* schema.

The *SKUP* contains the elements $S = (s^0, s^1)$; $K = (k^0, k^1)$; $U = (u^0, u^1)$; and $P = (p^0, p^1)$.

Typical notations for conditions are O1, O2, O3, and O4 corresponding to the level $m = 0$ and O11, O12, and so on corresponding to the level $m = 1$.

The EDOE presentation is limited to O1 and O3 detailed study.

Let $O1 = k_0^0$, $O2 = k_1^0$, $O3 = k_2^0$ and $O4 = k_3^0$.

The evolution at the level $m = 0$ is represented by a trajectory of high-thickness border cells. The system initial state is s_0^0.

With possibility $p^0(k_0^0|s_0^0)$, the condition k_0^0 is selected.

This is the digit symbolizing a specific cost evaluation. Based on this, the operator u^0 defined by $u^0(k_0^0, s_0^0) = s_1^0$ allows the transition to the new state s_1^0. Then, with possibility $p^0(k_1^0|s_1^0)$, the new condition k_1^0 arises. This condition symbolized by a digit corresponds to the selection of processes. Based on the operator $u^0(k_1^0, s_1^0) = s_2^0$, the system reaches the state s_2^0 and so on.

Observe that $s_1^0 = u^0(k_0^0, s_0^0)$ implies $s_2^0 = u^0(k_1^0, u^0(k_0^0, s_0^0))$ and so on.

This type of recursive relation emphasizes the role of the initial states and that of the conditions. The associated circular schema is $(s_0^0, k_0^0) \rightarrow (s_1^0, k_1^0) \rightarrow (s_2^0, k_2^0) \rightarrow (s_3^0, k_3^0) \rightarrow (s_0^0, k_0^0)$.

If simulation shows that $O1$ and $O3$ are significant factors, the analysis may be continued at the level $m = 1$ for the test conditions $O11 = k_0^1$, $O12 = k_1^1$, $O13 = k_2^1$, $O14 = k_3^1$, $O31 = k_4^1$, $O32 = k_5^1$, $O33 = k_6^1$, and $O34 = k_7^1$. The trajectory at the level $m = 1$ is illustrated by medium-thickness border cells. The system initial state at the level $m = 1$ is s_0^1. With possibility $p^1(k_0^1|s_0^1)$, the condition k_0^1 arises. This is a digit symbolizing $O11$, that is, the location of similar product. Based on this, the operator $u^1(k_0^1, s_0^1) = s_1^1$ describes the transition to the new state s_1^1 and so on.

With possibility $p^1(k_1^1|s_1^1)$, the condition k_1^1 arises. This is a digit symbolizing $O12$, that is, the step costs. Based on this, the operator $u^1(k_1^1, s_1^1) = s_2^1$ describes the transition to the new state s_2^1.

Observe that $s_2^1 = u^1(k_1^1, u^1(k_0^1, s_0^1))$.

The process continues a finite number of steps, as shown in Figure 16.1.

Figure 16.1 outlines the fractal nature of the framework. The high-thickness border cells loop corresponding to the level $m = 0$ generates smaller loops of medium-thickness border cells showing in fact the same structure at a different scale. These standardize the calculations. Moreover, this justifies the fractal manufacturing implementation.

16.7
Multiagent Manufacturing Systems: Architecture

An agent may be defined as a device or a self-directed program object that has its own value system and the means to solve certain tasks independently and communicate its solution to a larger problem-solving organization.

The main types of agents are

- Autonomous agents capable of effective independent actions
- Objective directed agents, when autonomous actions are directed toward the achievement of defined tasks
- Intelligent agents with the ability to learn and adapt
- Cooperative agents assisting other agents to perform a task.

Examples of multiagent systems are neurons in the brain, antibodies in case of immune systems, ants in colonies, wolfs in packs, investors in the stock market, people in social networks, and so on. In each case, agents have relatively limited set of rules, and the complexity of the collective behavior emerges from the large number of interactions they have among each other and their environment. There is constant action and reaction to what other agents are doing, thus nothing in the complex system is essentially fixed.

The multiagent architecture is significant for system functioning.

The organizations that are able to produce evolvable products would be ultimately evolvable ones (Hogg and Huberman, 1998; Ueda *et al.*, 2001). Modern organizations are facing complexity, chaos, interdependency, and interactions within and outside the boundaries. Global companies have begun to reorganize and decentralize their large structures to enable successful strategies despite increasing complexity of the production problems. They have started to support organizational fluidity with new pathways that allow project-centered groups to form rapidly and reconfigure as circumstances demand.

An open problem is how to structure the way that agents collaborate in complex conditions. Companies are looking for a system of cooperation among groups of agents that will work significantly better than individual groups. A group contains agents working together with a common purpose. It should be partial specialization but should also focus on specific tasks without losing the general objective. An important issue is how the structure and the fluidity of a group will affect the global dynamics of cooperation. Fluidity depends on how easily individual agents and information can move within the company structure and how easily they can break away on their own, extending the structure. The capability to manage the complexity of modern companies depends decisively of an effective communication network.

Agility in global companies is no longer measured only by the flexibility and responsiveness of a single center but also by the agility of the network of several centers and conditioning levels.

To compare different organizations, an informational entropy H calculus based on similarities as a measure of communication degree can be done. The entropy is defined by Equation 5.11. The similarity between the groups is $(0.5)^k$, where k is the number of steps till the hierarchical level at which their communication is possible. The similarity of a group with itself is 1. In this case, $k = 0$. If the groups need to move two higher levels up for communication, the similarity is 0.125. In this case, $k = 2$.

Some types of groups or individual organizations of interest are presented in Figure 16.2.

The illustrative examples are

(a) Isolated groups, $H = 0$
(b) Local adjacent groups, $H = 2.77$
(c) Hierarchical tree organizations, $H = 7.27$
(d) Hierarchical with horizontal associations, $H = 6.146$
(e) Multihierarchical groups, $H = 7.53$
(f) Multicentric groups, $H = 8.31$.

For local organization cases, the communication is only inside the same group (Figure 16.2a) or between the adjacent groups (Figure 16.2b). The proximity makes two elements to be grouped together. This does not ensure real communication between all the groups. The informational entropy is zero or very low. The entropy is $H = 0$ for the diagram shown in Figure 16.2a and $H = 2.77$ for Figure 16.2b.

Hierarchical organizations are shown in Figure 16.2c. The activity is performed at multiple levels and scales to achieve local and global issues. In the example presented in the figure, there are four groups managed by two managerial groups, directed by one directorial group. Managerial groups of different levels handle the information. The hierarchical pattern of communication is based on the following rule: the groups

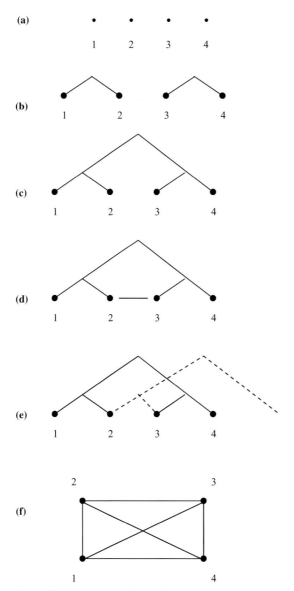

Figure 16.2 Organizations for multiagents systems.

communicate to the managerial group that is responsible for them and the managerial groups communicate to the directorial group. This allows control decisions to be
made at various scales, while maintaining local responsiveness and limiting the
required communication. The system has low complexity but relatively high informational entropy. In this type of hierarchical architecture, a complex problem is

decomposed in several simpler and smaller problems, and distributed among multiple control layers. This architecture allows the distribution of decision-making among the hierarchical levels. The main advantages are the robustness, the predictability, and the efficiency. However, the appearance of disturbances in the system reduces significantly its performances.

The modified hierarchical architecture tries to find a solution to the reaction-to-disturbance problems, maintaining all the features of hierarchical architectures and adding the interaction between modules at the same hierarchical level. This interaction allows the exchange of information between the modules and improves the reaction to disturbances. Hierarchical organizations with horizontal associations are shown in Figure 16.2d. In this case, the groups establish supplementary communication patterns and the system becomes less regular and more complex. For the case presented in Figure 16.2d, one horizontal association between group 2 and 3 is included. The complexity of the system increases, but the system loses entropy.

Multihierarchical organization is shown in Figure 16.2e. This consists of a collection of overlapping hierarchies in which some groups pertain to more hierarchies. Groups will overlap as some individuals can be members of several informational hierarchies. It is expected that this is better than a hierarchy in achieving stability, while having, in addition, a position invariant response that allows the control of disturbances at appropriate scale and location (Huberman and Hogg, 1986; Hogg and Huberman, 1998).

In the case of the multicentric interactive net (Figure 16.2f), the complete information is available to different groups making control decisions. This is not easy to do in practice for large-scale systems. These organizations allow a high performance against disturbances, but the global optimization is reduced because decision-making is local and autonomous, without a global view of the system. For these reasons, a center is necessary.

A better organization would consist of groups self-organized in concentric levels around the center. The information is changed easily for groups pertaining to the same level and passes through different other levels. More velocities and timescales, more sequences, for information transfer have to be considered. Teamwork may result from the activity of small groups with shared purposes, diversity of skills, coherence, and communication. Complex cooperative behavior is spontaneously emergent provided that the groups are small, diverse in composition, have long outlooks and fluid structure, and a pattern of interdependencies that varies. The interactive multicentric net in continuous and recurrent reorganization movement represents a potential model for the future evolvable organizations.

16.8
Multiagent-Based Versus Conventional Manufacturing Systems

The distributed manufacturing environments and the flexibility and reaction to disturbances requirements are crucial reasons for moving to new organizational paradigms.

Next generation of manufacturing control systems comprises the high adaptation and reaction to the occurrence of disturbances and environment changes. On the contrary, these control systems should optimize the global performance of the system that requires a global view of the entire system. These requirements imply the development of new manufacturing control systems with more autonomy, robustness against disturbances, and ability to handle changes and disturbances much better than the actual systems. The new paradigms should focus on the ability to respond promptly and correctly to external changes, without external interventions. Distributed manufacturing architectures, multiagent-based manufacturing systems seem to respond to complexity challenges (Parunak and Brueckner, 2001).

Table 16.1 compares the multiagent-based manufacturing systems with conventional hierarchical approaches.

The autonomous multiagent systems may have some disadvantages. Theoretical optima cannot be guaranteed. Predictions can be made only at the aggregate level. Systems of autonomous agents can become computationally unstable.

However, an autonomous approach appears to offer some significant advantages over conventional systems. Because each agent is close to the point of contact with the real world, the system computational state tracks the state of the world closely, without the need for a centralized database.

Multiagent systems offer a way to relax the constraints of centralized, planned, and sequential control. They offer production systems that are decentralized rather than centralized, emergent rather than planned, and concurrent rather than sequential.

The *SKUP* quadruple elements are linked to multiagent systems frames.

In the multiagent-based simulation described by Edmonds (2000), the elements of *K* are inferences using the model, the prediction/decoding corresponds to operators *U*, the natural process to the states *S*, and the measuring/encoding to possibilities *P*.

For the Parunak and Brueckner (2001) study of coordination, the agents with their specific roles represent the elements of *K*, the so-called rational actions are naturally linked to the elements of operators *U*, the perceptions are elements of possibilities *P*,

Table 16.1 Multiagent-based versus conventional manufacturing systems.

Characteristics	Conventional	Multiagent systems
Model source	Military	Biology, economy, and society
Optimum	Yes	No
Prediction level	Individual	Aggregate
Computational stability	High	Low
Match to reality	Low	High
Requires central data	Yes	No
Response to change	Fragile	Relatively robust
System reconfiguration	Hard	Easy
Calculus	Complicated, long	Simple, short
Time required for schedule	Slow	Real time
Processing	Sequential	Concurrent, parallel

while the states S are represented in this particular case by pheromones. The elements of the set K, that is, the agents, may be considered as structured too.

A typical conceptual structure of an agent includes objectives, knowledge processor, preceptor, actuator, and communication parts.

In the categorical frame, the perception P and the action U have been viewed as adjoint functors.

The objective consists of a list of roles for an agent to play. The knowledge processor is a knowledge base system that stores and processes the necessary knowledge for an agent to play the role designed for it by the agent set. Perception is a channel for an agent to receive information from the external world.

Actuator is an interface for an agent to modify or influence the states of an agent community.

These are completed by a communication mechanism for an agent to exchange views with other members in the agent set.

The switch from categorical product to coproduct controls the structure of the agent population and allows the emergence of new objectives and tasks. This circle of ideas has been studied as cognitive/reactive strategy in multiagent systems. The cognitive or interactive step, correlated to categorical product, implies creating new solutions. The reactive step, correlated to coproduct, selects and links solutions for further development.

Notice that the multiagent approach is suited for applications of other manufacturing paradigms. Multiagent technology provides techniques for modeling and implementing autonomous and cooperative software systems and represents an enabling technology for control of holonic manufacturing systems. Agents can be viewed as holons without physical processing capabilities.

The multiagent technology was useful in implementing fractal manufacturing systems that can reconfigure with respect to system environment (Ryu, Son and Jung, 2003).

Part Five
Concluding Remarks

Evolvable Designs of Experiments: Applications for Circuits. Octavian Iordache
Copyright © 2009 WILEY-VCH Verlag GmbH & Co. KGaA, Weinheim
ISBN: 978-3-527-32424-8

17
Related Concepts

At the end of this incursion in complex problem solving domain, it is appropriate to reinforce the connection of evolvable designs with some other theoretical concepts and research directions. This will help in understanding where future studies will be heading and in promoting evolvable design utilization for practical purposes.

17.1
Complexity

Recent technology developments and market requests have led to a strong increase in the complexity of industrial systems.

Complex systems are commonly understood as multilevel systems resulting from nonlinear interactions of several interdependent parts and displaying emergent behavior. Emergent phenomena arise as an outcome of the interactions between the constituent units of a system. Emergent phenomena are difficult to predict or percept, because they are properties of the whole that cannot be reduced to the behavior of the parts.

Despite the fact that numerous engineering systems are recognized as complexes, more of the traditional ones continue to be operated in regimes, where complexity properties can be neglected. The challenge for engineers and industrialists is not only to identify but also to cross the complexity barrier.

Confronted with inefficiency of reductionism, several companies lost confidence in predictive and theoretical methods and made decision mainly by expensive experiences. This may be the cause of decline in productivity and competitiveness. To regain industry confidence, there is a need to develop innovative methodologies based on an integrated approach that takes the complexity into account. It is a request for a future engineering science that will help produce high complexity products. It should be emphasized that complexity science is still evolving and is the object of intensive research all around the world.

Evolvable Designs of Experiments: Applications for Circuits. Octavian Iordache
Copyright © 2009 WILEY-VCH Verlag GmbH & Co. KGaA, Weinheim
ISBN: 978-3-527-32424-8

17.2
Evolvability

A paradigm shift from optimization to evolvability is evaluated in this book.

It was observed that the more an autonomous system is optimized to perform a task, the less capability it has to deal with unexpected changes in an uncertain environment. Consequently, for complexity conditions, the evolvability rather than optimization is the suitable measure of the potential to solve problems. In numerous practical situations, it may be impossible to achieve the optimum owing to the restrictions for time and resources. Complex system studies are grounded on evolvability rather than optimization criteria. Anything that manages to evolve will be adapted to the environment in which it happens to be.

Only evolvability permits to confront and break the complexity barrier. The emergence is linked to the concept of evolvability, since the evolution is considered as the mechanism for emergence.

Evolvable designs of experiments (EDOEs) are presented as an approach to meet the requirements imposed on the problem solving methods by complexity. EDOEs are networks of component designs of experiments that result from the evolvable self-referencing autonomous closure between the symbolic rules of the designs of experiments (DOEs) and the dynamics of the real physical organization.

17.3
Polystochastic Method

The key mathematical tool for EDOE modeling are the polystochastic models (PSMs).

PSM describe the systems that emerge when several component stochastic processes going on at different conditioning levels are able to interplay with each other, resulting in a process that is structurally different from the component processes.

PSM and the associated cognitive architecture ($SKUP$: S, states; K, conditions; U, operators; and P, possibilities) are the tools provided to study complexity and evolvability.

The $SKUP$ outlines the comprehensive architecture shared by the scientific or engineering method and also by the operational structure of evolvable devices, and the functional organization of organisms as informational and cognitive systems. The $SKUP$ outlines two sets (categories) S and K, and two transforming functions (functors) U and P in between these sets.

Specificities and claims of the methodology proposed include the following:

- Complementary algebraic frames for states-S (dynamical, natural, and real) and conditions-K (symbolic, formal, and "other than real").
- Several levels and scales for $SKUP$ elements.
- Differential model for the K process.
- Possibilities P, supplementing probabilities, to express potentiality, fuzziness, embedding, classification, and emergence.
- Informational entropy and informational distance as criteria for evolvability.
- Constructivist approach.

Compact wave equation models and *SKUP* method allow to downgrade the complex problem solving to Cartesian-inspired approaches as, for instance, that offered by the troubleshooting guide presented in Chapter 12. Such decision tables contain if–then rules applied linearly, for instance, if that failure signature is observed then that corrective action should be performed.

17.4
Constructivism

The run through complexity frontier in problem solving needs a constructivist approach, that is, the replacement of the preprogrammed and fixed DOE by the actively constructed and evolvable DOE.

Constructivism is based on the thesis that knowledge cannot be a passive reflection of reality, or a passive application of a formal model, but has to be more of an active and interactive construction. This view makes use of the notion of scheme as a collection of actions and thoughts that agents use to interact with the world and to solve problems. According to Piaget (1971), only by assimilation and accommodation, the cognitive agent constructs a collection of rules and schemes during the ontogeny. Knowledge is defined by schemes that are regarded true, but are ready to be changed. Constructivism is considered as a means to forgo the rigidity that prevents science from becoming more productive than it is today. Evolvability criterion explains why industrialists cannot construct any reality they would like to.

17.5
Cybernetics and the Cycle of Sciences

The cybernetic approach will be correlated to the cycle of sciences as described by Piaget (1967). This clarifies the position of EDOE in complexity study.

The traditional view on classification of systems and associated sciences is a linear and hierarchical one. Piaget (1967) disagreed with this model. The sciences are related cyclically since there is none to look up to for logical and mathematical phenomena other than in psychological activity.

According to Piaget, the circular model goes as follows: psychological phenomena depend on biological phenomena that in turn depend on physical and chemical phenomena that in their turn are stated in mathematical laws, and with these, we are back at the beginning, namely, at psychological phenomena (Table 17.1).

Traditionally, the field of cybernetics has been described as being developed in two steps, namely, the first- and second-order cybernetics.

The first-order cybernetics is the science of control and communication for machines and animals outlining the feedback concept. It focused on the idea of homeostasis, the ability of systems to maintain steady states despite disturbances in the environment. The applications are interdisciplinary between the sciences of mater and the biosciences.

Table 17.1 Cycle of sciences and cybernetics.

Biosystems Biology Anatomy	⟷Second-order reflexivity, self-organization	Cognitive Engineering Design Psychology Sociology
↕ First-order homeostasis, feedback		↕ Third-order virtual, anticipation
Material Physics Chemistry	⟷Fourth-order embodiment, evolvability	Mathematics Logics Philosophy

The second phase in the development of cybernetics focused on incorporating reflexivity into the systems, that is, acknowledging the observer to be a part of the observed system. This led to a more serious study of reflexivity and self-organization concepts. The applications of the second-order cybernetics are interdisciplinary between biosciences and cognitive sciences.

Possible third- and fourth-order cybernetics will confront the modern challenge of complexity arriving in science and technology.

The idea of emergence is fundamental here – the thought that complex systems, when recursively structured, can spontaneously evolve in directions their designers did not anticipate. It is the case of proactive devices. The virtual is a significant characteristic of third-order cybernetics. In third-order cybernetics, the system can change goals without preprogramming. This means that the observer is considered as a proactive component that not only observes but also decides and acts. The observer is not necessarily a human.

Third-order cybernetics elements may be found in the study of the third-wave cybernetics (Hayles, 1999), the study of viable systems (Yolles and Dubois, 2001), the study of social systems, and of cybersemiotics.

The applications of third-order cybernetics are beyond cognitive sciences and include virtual, proactive, anticipative technologies, and cyberspace. The third wave focuses on virtual systems, on building information systems capable of evolving.

Fourth-order cybernetics confronts and surpasses the challenge of complexity arriving in technology and life.

The fourth-order cybernetics may be one of the embodied, fully evolvable of the creative systems. This kind of cybernetics seems to be closely related to artificial life domain.

Artificial life is a study of life as it could be. One of its challenges is to construct living technologies or evolvable devices (Bedau *et al.*, 2000).

It should be noted that any new type of cybernetics embeds the elements of the previous ones. Since, after one cycle, the material embodiment of mathematics and computing capacity may support the emergence of another type of sciences of matter, a spiral of sciences instead of cycle of sciences and associated systems may be considered.

References

Abelson, H., Allen, D., Coore, D., Hanson, C., Homsy, G., Knight, T., Nagpal, R., Rauch, E., Sussman, G. and Weiss, R. (2000) *Communications of the ACM*, **43**, 74–82.

Adami, C. (2002) *Bioessays: News and Reviews in Molecular, Cellular and Developmental Biology*, **24**, 1085–1094.

Adleman, L.M. (1994) *Science*, **266**, 1021–1024.

Albus, J. (1991) *IEEE Transactions on Systems, Man and Cybernetics*, **21**, 473–509.

Albus, J. and Meystel, A. (2001) *Engineering of Mind: An Introduction to the Science of Intelligent Systems*, John Wiley & Sons, Inc., New York.

Altshuller, G. (1984) *Creativity as an Exact Science*, Gordon and Breach, New York.

Andrews, G., Birney, D. and Halford, G.S. (2006) *Memory & Cognition*, **34**, 1325–1340.

Austin, J. (1995) *International Journal on Fuzzy Sets and System*, **82**, 223–233.

Banzhaf, W., Dittrich, P. and Rauhe, H. (1996) *Nanotechnology*, **7**, 307–314.

Barnsley, M.F. (1993) *Fractals Everywhere*, Academic Press, New York.

Barr, M. and Wells, C. (1995) *Category Theory for Computing Science*, Prentice Hall, New York.

Bedau, M.A., McCaskill, J.S., Packard, N.H., Rasmussen, S., Adami, C., Green, D.G., Ikegami, T., Kaneko, K. and Ray, T.S. (2000) *Artificial Life*, **6**, 363–376.

Bird, J. and Layzell, P. (2002) The evolved radio and its implications for modeling the evolution of novel sensors. Proceedings of the Congress on Evolutionary Computation, CEC2002, pp. 1836–1841.

Birge, R. (1995) *Scientific American*, **272**, 90–95.

Bochmann, D. and Posthoff, C. (1981) *Binare Dynamische Systeme*, Akademieverlag, Berlin.

Boschetti, F., Prokopenko, M., Macreadie, I. and Grisogono, A. (2005) Defining and detecting emergence in complex networks. Knowledge-Based Intelligent Information and Engineering Systems, 9th International Conference, KES 2005, Melbourne, Australia, September 14–16, 2005, Proceedings, Part IV, Lecture Notes in Computer Science, **vol. 3684** (eds R. Khosla, R.J. Howlett and L.C. Jain), pp. 573–580.

Box, G.E.P., Hunter, W.G. and Hunter, J.S. (1978) *Statistics for Experimenters – Introduction to Design, Data Analysis and Model Building*, John Wiley & Sons, Inc., New York.

Carbone, A. and Seeman, N.C. (2002a) *PNAS*, **99**, 12577–12582.

Carbone, A. and Seeman, N.C. (2002b) *Natural Computing*, **1**, 469–480.

Cariani, P. On the Design of Devices with Emergent Semantic Functions, Ph.D. Dissertation, Binghamton University (1989).

Cariani, P. (1993) *Systems Research*, **10**, 19–33.

Cariani, P. (2001) *Bio Systems*, **60**, 59–83.

Chaput, H.H., Kuipers, B. and Miikkulainen, R. (2003) Constructivist learning: a neural implementation of the schema mechanism. Proceedings of the WSOM'03, The Workshop of Self-Organizing Maps, Kitakyushu, Japan.

Cheung, P., Berlin, A., Biegelsen, D. and Jackson, W. (1997) Batch fabrication of pneumatic valve arrays by combining MEMS

Evolvable Designs of Experiments: Applications for Circuits. Octavian Iordache
Copyright © 2009 WILEY-VCH Verlag GmbH & Co. KGaA, Weinheim
ISBN: 978-3-527-32424-8

with PCB technology. Symposium on Micro-Mechanical Systems, ASME, Dallas, TX, **62**, pp. 39–46.

Chomsky, N. (1966) *Cartesian Linguistics*, Harper & Row, New York.

Cohen, J.E. (1979) *Stochastic Processes and their Applications*, **9**, 245–251.

Condra, L.W. (1993) *Reliability Improvement with Design of Experiment*, Marcel Dekker, New York.

Cook, M., Rothemund, P.W.K. and Winfree, E. (2004) Self-assembled circuit patterns, in *DNA Computing* (eds J. Chen and J. Reif), 9 LNCS, 2943, Springer, Berlin, pp. 91–107.

Coombs, C.F. (1996) *Printed Circuits Handbook*, Mc-Graw Hill, New Jersey.

Deming, W.E. (1982) *Quality, Productivity, and Competitive Position*, MIT Press, Cambridge, MA.

Dewar, R.C. (2003) *Journal of Physics A: Mathematical and General*, **36**, 631–641.

Dey, A. (1985) *Orthogonal Fractional Factorial Designs*, John Wiley & Sons, Inc., New York.

Dini, J.W. (1993) *Electrodeposition. The Materials Science of Coating and Substrates*, Noyes Publication, Park Ridge, IL.

Dittrich, P., Ziegler, J. and Banzhaf, W. (2001) *Artificial Life*, **7**, 225–275.

Drescher, G. (1991) *Made-Up Minds: A Constructivist Approach to Artificial Intelligence*, MIT Press, Cambridge, MA.

Dubois, D. and Prade, H. (2001) *Annals of Mathematics and Artificial Intelligence*, **32**, 35–66.

Eden, C. (1988) *European Journal of Operational Research*, **36**, 1–13.

Edmonds, B. (2000) The use of models-making MABS actually work, in *Multi Agent Based Simulation* (eds S. Moss and P. Davidsson), Lecture Notes in Artificial Intelligence, LNAI, 1979, Springer, Berlin, pp. 15–32.

Elliot, F.D. and Rao, K. (1982) *Fast Transforms*, Academic Press, New York.

Erickson, D. and Li, D. (2004) *Analytica Chimica Acta*, **507**, 11–26.

Findley, G.L., Findley, A.M. and McGlynn, S.P. (1982) *PNAS*, **79**, 7061–7065.

Fodor, J. and Pylyshyn, Z. (1988) *Cognition*, **28**, 3–71.

Ganter, B. and Wille, R. (1999) *Formal Concept Analysis. Mathematical Foundations*, Springer, Berlin.

Giry, M. (1981) A categorical approach to probability theory, in *Categorical Aspects of Topology and Analysis* (ed. B. Banaschewski), Lecture Notes in Mathematics, 915, Springer, Berlin, pp. 68–85.

von Glasersfeld, E. (1995) *Radical Constructivism: A Way of Knowing and Learning*, Falmer Press, London.

Goranovic, G., Rasmussen, S. and Nielsen, P.E. (2006) Artificial life forms in microfluidic systems. Proceedings microTAS 06, Tokyo, Japan, 2, p. 1408.

Goulet, M. (1992) *Bare Board Drilling*, Miller Freeman, San Francisco, CA.

Halford, G.S., Wilson, W.H. and Phillips, S. (1998) *Brain and Behavioural Sciences*, **21**, 803–864.

Harmuth, H.F. (1977) *Sequence Theory, Foundations and Applications*, Academic Press, New York.

Haronian, D. and Lewis, A. (1991) *Applied Optics*, **30**, 597–608.

Hayles, K. (1999) *How We Became Posthumans. Virtual Bodies in Cybernetics, Literature and Informatics*, Chicago UP, Chicago.

Hedayat, A.S., Sloane, N.J.A. and Stufken, J. (1999) *Orthogonal Arrays. Theory and Applications*, Springer, New York.

Hedayat, A.S. and Stufken, J. (1999) *Technometrics*, **41**, 57–61.

Hersh, R. (2003) *The Mathematical Intelligencer*, **25**, 53–60.

Hjelmfelt, A., Schneider, F.W. and Ross, J. (1993) *Science*, **260**, 335–337.

Hogg, T. and Huberman, B.A. (1998) *Smart Structures and Materials*, **7**, R1–R14.

Horgan, J. (1995) *Scientific American*, **272**, 104–109.

Horowitz, S.J. and Needes, C.R.S. (1997) *Future Circuits International*, **1**, 25–30.

Huberman, H.A. and Hogg, T. (1986) *Physica D*, **2**, 376–384.

Hummel, J.E. and Choplin, J.M. (2000) Toward an integrated account of reflexive and reflective reasoning. in *Proceedings of the Twenty Second Annual Conference of the*

Cognitive Science Society (eds L.R. Gleitman and A.K. Joshi), LEA, Mahwah, NJ, pp. 232–237

Hummel, J.E. and Holyoak, K.J. (1997) *Psychological Review*, **104**, 427–466.

Hwang, J.S. (1996) *Modern Solder Technology for Competitive Electronics Manufacturing*, McGraw-Hill, New York.

Inhelder, B. and Piaget, J. (1958) *The Growth of Logical Thinking from Childhood to Adolescence*, Basic Books, New York.

Iordache, O. (1987) *Polystochastic Models in Chemical Engineering*, VNU Science Press, Utrecht, The Netherlands.

Iordache, O. and Corbu, S. (1987) *Chemical Engineering Science*, **42**, 125–132.

Iordache, O., Bucurescu, I. and Pascu, A. (1990) *Journal of Mathematical Analysis and Applications*, **146**, 306–317.

Iordache, O. (1992) Dyadic frames for intermittency. Perturbed models. in *Proceedings of the Conference on p-adic Functional Analysis, Laredo (Spain), May 1990* (eds J.M. Bayod, N. De Grande De-Kimpe and W.H. Schikhof), Lecture Notes in Pure and Applied Mathematics, 137, Marcel Dekker, New York, pp. 89–99

Iordache, O., Corriou, J.P., Garrido-Sanchez, L., Fonteix, C. and Tondeur, D. (1993a) *Computers and Chemical Engineering*, **17**, 1101–1113.

Iordache, O., Valentin, G., Corriou, J.P., Pons, M.N. and Pethö, A. (1993b) *Acta Chemica Hungarica, Models in Chemistry*. **130**, 1–18.

Iordache, O., Corriou, J.P. and Tondeur, D. (1993c) *Canadian Journal of Chemical Engineering*, **71**, 955–966.

Iordache, O. (1996) *Material Science and Engineering C*, **4**, 143–148.

Iordache, O. and St. Martin, S. (1996) Quality evaluation for direct metallization. Proceedings IPC Expo Meeting, San Jose, March, 1996, pp. S13-2.1–S13-2.16.

Iordache, O. and St. Martin, S. (1997) Test coupons for PCB quality control and evaluation. Proceedings IPC Expo Meeting, San Jose, March, 1997, pp. S8-3.1–S8-3.9.

Iordache, O. (1998a) Can accelerated stress tests be "Lingua Franca" for reliability tests?

Proceedings European IPC, EIPC Convention, Wiesbaden, Sept., 1998, pp. 7-2.1–7-2.17.

Iordache, O. (1998b) *Future Circuits International*, **4**, 161–168.

Iordache, O., Jianxiong, Hu and St. Martin, S. (1998a) Solderability and surface finishes: evaluation and building. Proceedings IPC Expo Meeting, Long Beach, April 1998, pp. S11-4–S11-4.12.

Iordache, O., Jianxiong, Hu and St. Martin, S. (1998b) Drilling quality for microvia. in Proceedings IPC Expo Meeting, Long Beach, April, 1998, pp. S19-5.1–S19-5.8.

Iordache, O. and St. Martin, S. (1998) *Future Circuits International*, **3**, 129–133.

Iordache, O. and Hu, J. (1999) Reliability tests correlation. "Lingua Franca" methodology. Proceedings IPC Expo Meeting, Long Beach, March, 1999, pp. S14-6.1–S14-6.14.

Iordache, O. and Jianxiong, Hu (2000a) Informed circuits. Accelerated failure analysis. Proceedings IPC Expo Meeting, San Diego April 2000, pp. S4-2.1–S4-2.10.

Iordache, O. and Jianxiong, Hu (2000b) *Future Circuits International*, **6**, 161–168.

Iosifescu, M. and Grigorescu, S. (1990) *Dependence with Complete Connections and Applications*, Cambridge University Press, Cambridge.

Iwata, K., Onosato, M. and Koike, M. (1994) *Annals of CIRP*, **43**, 379–383.

Joyal, A., Street, R.H. and Verity, D. (1996) *Mathematical Proceedings of the Cambridge Philosophical Society*, **119**, 447–468.

Kaehr, R. and von Goldammer, E. (1988) *Journal of Molecular Electronics*, **4**, 31–37.

Kauffman, S. (1995) *At Home in the Universe: The Search for Laws of Self-Organization and Complexity*, Oxford University Press, Oxford.

Keepler, M. (1998) *Portugaliae Mathematica*, **55**, 391–400.

Kemeny, J.G. and Snell, J.L. (1960) *Finite Markov Chains*, Van Nostrand, New York.

Khuri, S. (1994) Walsh and Haar functions in genetic algorithms. Proceedings of the 1994 ACM Symposium of Applied Computing, ACM Press, New York, pp. 201–205.

Kidd, P. (1994) *Agile Manufacturing, Forging New Frontiers*, Addison-Wesley, Boston, MA.

Klir, G.J. and Bo, Y. (1995) *Fuzzy Sets and Fuzzy Logic: Theory and Applications*, Prentice Hall, Englewood Cliffs, NJ.

Koza, J.R. (1999) *Genetic Programming III*, Morgan Kaufmann, San Fransisco, CA.

Kugler, P.N. and Turvey, M.T. (1987) *Information, Natural Law and the Self-Assembly of Rhytmic Movement*, Lawrence Erlbaum, Hillsdale, NJ.

Kuhnert, L., Algadze, K.I. and Krinsky, V.I. (1989) *Nature*, **337**, 244–247.

Lang, W.C. (1998) *Houston Journal of Mathematics*, **24**, 533–544.

Lehn, J.M. (2004) *Reports on Progress in Physics*, **67**, 249–265.

Lochner, R.H. and Matar, J.E. (1990) *Designing for Quality*, ASQC Quality Press, Milwaukee, WI.

Lucas, J.M. (1994) *Journal of Quality Technology*, **26**, 248–260.

Lucca, J., Sharda, R. and Weiser, M. (2000) Coordinating technologies for knowledge management in virtual organizations. Proceedings of the Academia/Industry Working Conference on Research Challenges, 2000, Buffalo, NY.

Luger, G.F., Lewis, J. and Stern, C. (2002) *Brain, Behavior and Evolution*, **59**, 87–100.

Macias, N.J. (1999) The PIG paradigm: the design and use of a massively parallel fine grained self-reconfigurable infinitely scalable architecture. Proceedings First NASA/DoD Workshop on Evolvable Hardware.

MacLane, S. (1971) *Categories for the Working Mathematician*, Springer, New York.

Mahalik, N.P. (2005) *Micromanufacturing and Nanotechnology*, Springer, New York.

Mainzer, K. (1996) *Thinking in complexity, the Complex Dynamics of Matter, Mind and Mankind*, Springer, Berlin.

Mange, D., Stauffer, A., Petraglio, E. and Tempesti, E. (2004) *Physica D*, **191**, 178–192.

Mann, S.(ed.) (1996) *Biomimetic Materials Chemistry*, Wiley-VCH Verlag GmbH, Weinheim, Germany.

Mansfield, E. (1988) *Management Science*, **34**, 1157–1168.

Maturana, H. and Varela, E. (1979) *Autopoiesis and Cognition*, Reidel, Boston, MA.

McCarthy, I.P., Rakotobe-Joel, T. and Frizelle, G. (2000) *Internation Journal of Manufacturing Technology and Management*, **2**, 559–579.

Miller, J.F. and Downing, K. (2002) Evolution in materio: looking beyond the silicon box. Proceedings of the 2002 NASA/DoD Conference on Evolvable Hardware, IEEE Computer Society Press, pp. 167–176.

Mnif, M. and Müller-Schloer, C. (2006) Quantitative emergence. Proceedings of the 2006 IEEE Mountain Workshop on Adaptive and Learning Systems (SMCals 2006), Pacificaway, NJ, pp. 78–84.

Montgomery, D.C. (1984) *Design and Analysis of Experiments*, John Wiley & Sons, Inc., New York.

Mori, T. (1995) *Taguchi Techniques for Images and Pattern Developing Technology*, Prentice Hall PTR, Englewood Cliffs, NJ.

Noort, D., van Wagler, P. and McCaskill, J.S. (2002) *Smart Materials & Structures*, **11**, 756–760.

Novak, J.D. (1991) *The Science Teacher*, **58**, 45–49.

Parunak, H.V.D. and Brueckner, S. (2001) Entropy and self-organization in multi-agent systems. Proceedings of the Fifth International Conference on Autonomous Agents, ACM Press, pp. 124–130.

Pattee, H.H. (1995) *Artificial Intelligence*, **12**, 9–28.

Pattee, H.H. (2000) Causation, control and the evolution of complexity, in *Downward Causation* (eds P.B. Anderson, P.V. Christiansen, C. Emmeche and N.O. Finnemann), Aarhus University Press, Aarhus, Denmark, pp. 63–67.

Phadke, M.S. (1989) *Quality Engineering Using Robust Design*, Mc-Graw Hill, New York.

Piaget, J. (1967) Classification des sciences et principaux courants épisté mologiques contemporains, in *Logique et connaissance scientifique* (ed. J. Piaget), Gallimard, Paris, 1151–1224.

Piaget, J. (1970) *Genetic Epistemology*, Columbia University Press, New York.

Piaget, J. (1971) *The construction of Reality in the Child*, Ballantine Books, New York.

Piaget, J. (1990) *Morphismes et Catégories*, Delachaux et Niestlé, Neuchatel, Switzerland.

Pierce, D. and Kuipers, B.J. (1997) *Artificial Intelligence*, **92**, 169–229.

Plate, T. (1995) *IEEE Transactions of Neural Networks*, **6**, 623–641.

Plate, T. (2000) *The International Journal of Knowledge engineering and Neural Networks*, **17**, 29–40.

Pollack, J.B. (1990) *Artificial Intelligence*, **46**, 77–105.

Prigogine, I. (1980) *From Being into Becoming*, W.H. Freeman, San Francisco, CA.

Quartz, S. and Sejnowski, T. (1997) *Behavioral and Brain Sciences*, **20**, 537–596.

Rammal, R., Toulouse, G. and Virasoro, M.A. (1986) *Reviews of Modern Physics*, **58**, 765–788.

Rasmussen, S., Chen, L., Deamer, D., Krakauer, D.C., Packard, N.H., Stadler, P.F. and Bedau, M.A. (2004) *Science*, **330**, 963–965.

Riegler, A. (2005) *Constructivist Foundations*, **1**, 1–8.

Rocha, L.M. (2001) *Bio Systems*, **60**, 95–121.

van Rooij, A.C.M. (1978) *Non-Archimedean Functional Analysis*, Marcel Dekker, New York.

Rosen, R. (1991) *Life itself: A Comprehensive Inquiry into the Nature, Origin and Fabrication of Life*, Columbia University Press, New York.

Rosenbaum, P.R. (1996) *Technometrics*, **38**, 354–364.

Ryu, K., Son, Y. and Jung, M. (2003) *Computers in Industry*, **52**, 161–182.

Schikhof, W.H. (1984) *Ultrametric Calculus*, Cambridge University Press, Cambridge.

Sekanina, L. (2004) *Evolvable Components: From Theory to Hardware Implementations*, Springer, Berlin.

Shastri, L. and Ajjanagadde, V. (1993) *Behavioral and Brain Sciences*, **16**, 417–493.

Siegmund, E., Heine, B. and Schulmeyer, P. (1990) *International Journal of Electronics*, **69**, 145–152.

Smith, M. (1993) *Neural Networks for Statistical Modeling*, Van Nostrand, Reinhold, New York.

Sowa, J.F. (2000) *Knowledge Representation: Logical, Philosophical and Computational Foundations*, Brooks-Cole, Pacific Grove, CA.

Taguchi, G. (1986) *Introduction to Quality Engineering: Design Quality into Products and Processes*, Asian Productivity Organization, Tokyo.

Tangen, U., Wagler, P.F., Chemnitz, S., Goranovic, G., Maeke, T. and McCaskill, J.S. (2006) *Complexus*, **3**, 48–57.

Taylor, T.J. (2002) Creativity in evolution: individuals, interactions and environments, in *Creative Evolutionary Systems* (eds P.J. Bentley and D.W. Corne), Morgan Kaufmann, San Fransisco, CA.

Teuscher, C., Mange, D., Stauffer, A. and Tempesti, G. (2003) *BioSystems*, **68**, 235–244.

Thompson, A. (1998) *Hardware Evolution: Automatic Design of Electronic Circuits in Reconfigurable Hardware by Artificial Evolution*, Springer, Berlin.

Thompson, A., Layzell, P. and Zebulum, R.S. (1999) *IEEE Transactions on Evolutionary Computation*, **3**, 167–196.

Tomiyama, T. and Yoshikawa, H. (1987) Extended general design theory, in *Design Theory for CAD* (eds H. Yoshikawa and E.A. Warman), North Holland, Amsterdam, The Netherlands, pp. 95–130.

Ueda, K., Marcus, A., Monostori, L., Kals, H.J.J. and Arai, T. (2001) *Annals of the CIRP*, **50**, 535–551.

von Uexküll, J. (1928) *Theoretische Biologie*, Suhrkamp, Frankfurt Main.

Upton, D. (1992) *Manufacturing Review*, **5**, 58–74.

Verpoorte, E. and de Rooij, N.F. (2003) *Proceedings of the IEEE*, **91**, 930–953.

Vsevolodov, N. (1998) *Biomolecular Electronics: An Introduction via Photosensitive Proteins*, Birkhaeuser, Boston, MA.

Wassink, R.J.K. (1989) *Soldering in Electronics*, Electrochemical Publications Ltd., Ayr, Scotland.

Watson, I. (1997) *Applying Case-Based Reasoning: Techniques for Enterprise Systems*, Morgan Kaufmann, San Francisco, CA.

Whitley, L.D. (1991) Fundamental principles of deception in genetic search, in *Foundations of*

Genetic Algorithms (ed. G. Rawlins), Morgan Kaufmann, San Mateo, CA.

Wiedermann, J. and van Leeuwen, J. (2002) *AI Communications*, **15**, 205–216.

Wilson, H.W. and Halford, G.S. (1994) Robustness of tensor product networks using distributed representations. Proceedings of the Fifth Australian Conference on Neural Networks (ACNN'94), pp. 258–261.

Winfree, E. (2000) *Journal of Biomolecular Structure & Dynamics*, **11**, 263–270.

Yanuschkevich, S. (1998) *Logic Differential Calculus in Multi-Valued Logic Design*, Technical University of Szczecin, Szczecin, Poland.

Yolles, M.I. and Dubois, D. (2001) *International Journal of Computing Anticipatory Systems*, **9**, 3–20.

Yoshikawa, H. (1989) *Annals of the CIRP*, **38**, 579–586.

Zauner, K.P. (2005) *Critical Reviews in Solid States and Material Sciences*, **30**, 33–69.

Index

Evolvable Designs of Experiments: Applications for Circuits. Octavian Iordache
Copyright © 2009 WILEY-VCH Verlag GmbH & Co. KGaA, Weinheim
ISBN: 978-3-527-32424-8